ENVIRONMENT AND SOCIETY
THE CONTEXT OF NORTH-EAST INDIA

By the same author:

1. Forest in the Life of the Khasis
2. Socio-Economic Dimensions of Forest among the Khasis

The Editor

Dr. Rekha M. Shangpliang is Associate Professor in the Department of Sociology, NEHU. Her areas of specialization are Sociology of Environment, Sociology of Family and Kinship and Sociology of Gender. She has to her credit two books – *Forest in the Life of the Khasis* and *Socio-Economic Dimension of Forest among the Khasis* as well as a number of articles on Matriliny, Gender and Kinship published in National and International journals.

ENVIRONMENT AND SOCIETY
The Context of North-East India

Edited by

Rekha M. Shangpliang

Department of Sociology

North Eastern Hill University, Shillong

CONCEPT PUBLISHING COMPANY PVT. LTD.

NEW DELHI - 110 059

ISBN-13: 978-93-88937-34-4

First Published 2020

Published and Printed by

Concept Since 1974

Concept Publishing Company Pvt. Ltd.
A/15-16, Commercial Block, Mohan Garden
New Delhi-110059 (India)
T: 25351460, 25351794, *F*: 091-11-25357109
E: publishing@conceptpub.com, W: www.conceptpub.com

Cataloging in Publication Data--Courtesy: D.K. Agencies (P) Ltd. <docinfo@dkagencies.com>

National Seminar on "Environment and Society: the Context of North-East India" (2016 : Shillong, India)
Environment and society : the context of North-East India /
edited by Rekha M. Shangpliang.
pages cm
Papers presented at a two-day National Seminar on "Environment and Society: the Context of North-East India", organized by the Department of Sociology, North Eastern Hill University on 14th and 15th of March, 2016.
Includes bibliographical references and index.
ISBN 9789388937344

1. Human ecology—India, Northeastern—Congresses. 2. Natural resources—India, Northeastern—Management—Congresses. 3. Tribes—India, Northeastern—Social life and customs—Congresses. 4. Environmental protection—India, Northeastern—Congresses. I. Shangpliang, Rekha M., editor. II. North Eastern Hill University. Department of Sociology, organizer. III. Title.

LCC GF662.N67S46 2016 | DDC 304.209541 23

Foreword

The Special Assistance Programme was initiated by the University Grants Commission in 1963 keeping in view the recommendations of the Education Commission to facilitate University Departments having potential in research and teaching. The programme is intended to encourage the pursuit of excellence and teamwork in advanced teaching and research. The Special Assistance Programme is a facility provided to select University Departments. The prerequisites are many, including sustained research on a focused goal and the recognition of research of a community of teachers within a University department.

The Department of Sociology, North-Eastern Hill University, since its inception in 1976 has focussed on providing space, faculty positions, teaching programmes and encouraged research on the region in keeping with the objective of the University. Among the more significant contribution of the Department to our understanding of North-East India is its focus on the environment and related issues. With its wealth and incredible variety of flora and fauna, blessed with spectacular geography and range of climate, and inhabited by numerous tribes and communities, the region has been the focus of attention for its distinct geographical and environmental position in South Asia.

It is encouraging that the Department of Sociology under the Special Assistance Programme and with the support of the ICSSR-NERC organized a seminar on "Environment and Society: The Context North-East India". The papers presented therein, after editing, comprise the contents of this useful volume. The authors of the papers published in this volume come from across the region and across the social sciences and humanities disciplines. They bring into this volume the conceptual understanding of the relationship of the environment and the tribes and communities inhabiting the hills and plains they call home; traditional knowledge systems; role of women in forest management; environment conservation; use of natural resources and their impact and a whole array of related issues.

The articles published in the volume should further stimulate ways to preserve the fragile environment both by traditional and modern means and identify areas within the region that have not been given the academic attention and policy intervention they require. I am confident the Department of Sociology will take the discussion of its concern for what is around us forward in synergy with stakeholders. Among others we look forward to reference of this collection of essay by teachers in case studies, by academia for reviews and by agencies involved in environment conservation and policy-making.

David R. Syiemlieh
Former Vice-Chancellor,
Rajiv Gandhi University
Former Member and Chairman, UPSC

Acknowledgements

The present book is a result of my long drawn interest in producing a volume that would cater to the needs of the students pursuing a course on Sociology of Environment and also for scholars interested in the field of Environment. It, therefore, would not have been possible for me to cover all the relevant topics had I not got the help and support of many who have extended immense help and support in holding the seminar on "Environment and Society: The Context of North-East India" and also in bringing out this volume.

Firstly, I wish to thank the Indian Council of Social Science Research (NERC), Shillong, Meghalaya for support of grants in organizing the seminar, the articles of which have been published in this volume. I am extremely thankful to all the contributors who have very promptly submitted the revised version of their full papers in time. The papers covered in this volume cover a vast range of environmental issues in the North-East.

I would be failing in my duty if I did not say a word of thanks to all my colleagues at the Department of Sociology, NEHU, Prof. C.L. Imchen, Prof. I.L. Aier, Prof. A.K. Nongkynrih, Prof. B. Panda and Yusuf Khan who have encouraged and helped me in many ways since the time of preparation of the seminar till the publication of this volume.

My sincerest thanks goes to my colleague and Co-ordinator of SAP-DSA I, Prof. D.V. Kumar for his help and encouragement at all times, and for always instilling in me the vision to work harder.

It goes without saying that for a woman like me who has multiple roles to play at home and work, my family plays an important role in times of need. Therefore, I take this opportunity to thank all my family members specially my father, Mr. K.K. Muktan who being a prolific writer himself has helped me through his valuable comments and suggestions to improve the quality of the volume. My husband Shem for always being there and my little daughter Abigail, who with her singing talents entertained me and gave me some time to relax.

Last but not the least, I am thankful to Concept Publishing Company (P) Ltd., New Delhi for undertaking the publication of this book.

Editorial Note

The significance of the physical environment as a formative factor in shaping human life and existence in a given habitat has become an important area of interest to social scientists in recent times. It has been commonly held that the environmental conditions stimulate human responses in their ideas and actions which in the long run results in the formation of social habits and customs. Living in harmony with nature has been an integral part of traditional and indigenous societies. To them nature has often been exemplified as 'cultural space' expressed through its invocation and representation in myths, ethics, beliefs and folkways. Traditional wisdom of the indigenous communities in particular has endowed them with skills and practices that are reflective of their deep sense of eco-consciousness. This leads one to the concept of 'holistic ecology' by which we understand how tribal and indigenous communities have lived in close symbiosis with nature and natural resources and they view themselves as an integral component of the eco-system. Here, the environment figures as functioning in a dual capacity – on the one hand we have the physical components such as climate, topography, animal and insect life, vegetation and soil which directly or indirectly shape human activity and productivity, and on the other, man alters the landscape, fells trees, erodes soil, builds dams etc. moulding the earth to meet his own needs. Moving firmly within such parameters of 'integrated viewpoint' of humans within the eco-system boundaries we can pose a few questions before us – "Can society exist apart from environment?" Or "Is environment expressive of society and culture?"

The North-East of India presents a dual picture of bio-diversity and socio-cultural diversity. The tribal communities in this region have maintained a close symbiotic relationship with their natural environment since time immemorial. They have maintained long traditions of depending on the natural resources of their own localities and preserving it besides having an extensive knowledge of the biota, the terrain, the waters, the climate and its seasonal changes, the plants and animals, their habitat preferences and local distribution, behaviour and uses. Such expertise and knowledge generally referred to as Traditional

Ecological Knowledge (TEK) accumulated on the basis of their experience and has been applied by the local community in their everyday life to obtain food, medicines, fuelwood and other necessities and in avoiding dangerous situations. Forests besides being the source of economic livelihood, have often been revered as the natural abode of spirits. This 'sacred' character of the forest finds expression in the customary practice of preserving sacred groves amongst many communities in the North-East, which are held with high esteem even today. The agro-biocultural links of the tribes and their natural environment has found expression in the various indigenous food habits and agricultural practices of the tribal communities of the region such as *jhum* cultivation, which has generated a new area of interest in marginalization of land, agro-bio-diversity, food security and nutrition, indigenous economies and livelihoods etc.

However, what is of grave concern in recent times is the way in which drastic changes in the environment are taking place. Today we find issues of 'development' and 'environment' and the 'politics of environment' have taken a new turn and acquired centre-stage amongst academicians, environmentalists, scientists, planners and policy-makers. This brings us to the new environmental debate on development options in contemporary India and the ecological limits to growth and development.

The subject of environment being one of the thrust areas, in both research and academic activities, the Department of Sociology, NEHU has in the past couple of years conducted a number of academic activities related to environment. A National level seminar entitled "Nature, Culture and the Tribes of North-East India" was held in the year 2008, followed by another National seminar entitled "Socio-cultural dimensions of Nature in the Context of North-East India". Besides holding seminars on environment related themes, the Department has taken initiatives to commemorate World Environment Day on June 5th every year by holding panel discussions and symposiums. The Department is at present floating a course on Sociology of Environment to the 4th Semester Post-Graduate students of the Department.

Taking into account the rich ecological and cultural diversity of North-East India, a two-day National seminar on "Environment and Society: The Context of North-East India" was organized by the Department of Sociology on the 14th and 15th of March, 2016. This was another endeavour by the Department and it was hoped that the seminar would develop a better understanding of the reciprocal interactions between the bio-physical world and the human social world and in the long run contribute towards a healthy respect for the ecological sensibilities of indigenous communities. It was against this backdrop that this seminar tried to focus on the linkages between culture and ecology.

The main *objectives* of the seminar were:

1. To have a conceptual and theoretical understanding of the relationship between tribes and their natural environment.
2. To explore the indigenous wisdom of conservation of land, forest, agro-ecology, women and natural resource management, indigenous economies and livelihood patterns, local food systems and wild edibles, and Traditional Knowledge System (TKS) of medicinal plants etc.
3. To examine the challenges and problems of the tribes of the region with regard to their access to natural resources such as landlessness, interventions, displacement, in the light of the existing environmental legislations and enforcement.

All the papers deliberated in the seminar ranged from the social and theoretical framework of ecology to wide variety of emerging environment related issues concerning the North-East such as sustainable use and conservation practices of land and forest management, agro-biology, local food systems, medicinal plants, local economies and livelihood patterns, community rights over land and forest, livelihood interventions, development issues, Government legislations *versus* customary laws with regard to land and forest, Common Property Resources (CPRs) and access to natural resources etc.

The present book is a compilation of 20 chapters presented at the seminar. The book is divided into 4 sections comprising of a wide range of chapters which deal with different themes on environment and society. The first section is entitled **Environment and Society: Some Theoretical Considerations**, which begins with the keynote address delivered by **Prof. Sekh Rahim Mondal.** Stating the importance of understanding and examining the relationship between environment and humans, he mentioned that such a discourse has been the pre-occupation of philosophers and scientists since historical times. In recent times study on environment and man relationship has taken a central theme and prime concern owing to the issues of anthropogenic global climate change, particularly from the middle of twentieth century. The early theory of environmental sociology/anthropology like Environmental

Determinism, directly addressed to the question of environment-man and climate-culture relationship has been resurrected time and again, because of its naivety, as a result sociological and anthropological discourses in present century is more in the forms of environmental derivatives, like ecology, disaster are interpretive in character.

Considering the deep relationship between environment and human civilization to the central stage, the 2014 report of the task force on global climate change, commissioned by American Anthropologies Association (AAA) invigorated anthropological and sociological interest in global climate change as its uneven impact affects the millions of people all-over the world. The current challenges stem from environment and cultural choices on the societal level and the value of long record of human civilization compelled us to decide our choices for future. It has now been realized from social scientific research that the local knowledge and traditional knowledge of the folk communities of developing countries have the significant potentials to act as solutions for emerging environmental problems.

The North-East India presents a dual character of bio-diversity as well as human and cultural diversity. The traditional wisdom of the indigenous communities of this place is the reflective of their deep sense of eco-consciousness, which needs to be properly documented, analysed and realized by the world community.

It is, therefore, pertinent to have a deeper conceptual and theoretical understanding of the relationship between natural environment and the folk communities, with particular reference to their eco-consciousness and environmental management strategies. To deal with this the most emerging concepts like deep ecology, authentic ecology, environmental citizenship and noosphere theory were discussed.

The keynote address was based on facts gathered from review of relevant literatures and empirical observations of the author in the field of nature culture studies. The speaker holds the view that nature-man relationship needs to be examined both from positivistic as well as from interpretative approaches.

B. Panda in his chapter "Environmental Movements in North-East India" presents the nature of environmental movements which have erupted in the North-East India within the broad theoretical framework of collective action and new social movement perspective and critically examines the issues associated with environmental movements in the region. He attempts to explore the strategies employed by the communities in North-East India particularly in Meghalaya to preserve and protect their environment where nature seems to play an indomitable role in their world-view. The chapter further argues that there is a need for a concerted effort involving multiple actors such as the state, the corporate, the voluntary sector and the public at large towards creating environmental awareness thereby creating a sustained approach to preserve the pristine ecology.

Chintamani Rout in his chapter "Indigenous People and their Development: Legal and Constitutional Safeguards" focuses on the International Movement of Indigenous People which has grown with extraordinary speed. Today, indigenous peoples have emerged as a distinct category in international human rights. Indigenous peoples around the world have sought recognition of their identities, their ways of life and their right to traditional lands and resources. In September 2007, the General Assembly of the United Nations has adopted "Declaration on the Rights of Indigenous peoples". Article 1 of the Declaration declares that all indigenous peoples have the right to the full enjoyment, as a collective or as individuals, of all human rights and fundamental freedoms as recognized in the Charter of the United Nations. Thus, the struggle of the world's indigenous peoples has reached a new height and the responsibility is upon the Nation-States to respect the rights of the indigenous peoples to maintain their identity, culture and interest to provide an environment to participate in social, economic and political life. This Research chapter examines in detail the rights of indigenous tribal people, their knowledge of conservation of land, forest, agro-ecology and natural resource management and their livelihood pattern etc. guaranteed under different statutory laws and the Constitution of India.

Sanjib Kakoty in his chapter entitled "Ecology, Sustainability and Embedded Belief Systems" focuses on the importance of maintaining ecological balance and sustainability with the prevalent economic model of limitless consumption. Inhabited as it were by numerous communities, tribes and sub-tribes, the North-East is predominantly inhabited by indigenous communities, designated as tribal. Tribal communities and indigenous religions the world over are accepted repositories of ancient knowledge and wisdom. Significantly, with the tribal communities, such knowledge systems are living traditions. Thus, such living knowledge systems could be an important source of guidance for modern man in his quest for sustainability. This chapter looks at some of the larger issues of the sustainability paradigm and also focuses on certain traditional practices of the North-Eastern communities.

The second section of the book is devoted to **Women and Natural Resource Management Strategies.** In recent years, the role of women in natural resource management has acquired a new dimension with the increasing focus on decentralization of community resources specially in developing nations. In recent times due to changes in human-nature relationship and increasing pressure on land and other resources, there is invisibility of women's livelihood strategies, marginalization of women's rights over NTFPs (Non-Timber Forest Products) and other resource base leading to "feminization of poverty". The first two chapters in this section focus on women as resource managers and the problems faced by them in the light of legislative policies and development process thereby limiting their access to forest and livelihood strategies.

Lucy Zehol in her chapter "Naga Women as Natural Resource Managers: Is Revival of Tradition Always the Best Option?" focused on rural Naga women as the one who are most directly dependent on natural resources for their immediate livelihood and food security. However, in resource management, there are differences between the roles and responsibilities of women and men due to the socially constructed division of labour. Gender approach to the environment policy should attempt to understand the differences that exist between women and men in households and between subgroups within communities. Besides, there is a need to understand power structures and relations, thus recognizing the rights and responsibilities of women and men that exist in communities. These may be partly hidden and may not be obvious without careful investigation. This chapter seeks to examine rural Naga Women's experiences as resource managers and how their experience help to raise some gender issues in resource management, which acts as a double-edged sword, focusing on the matrilineal Khasis of Meghalaya.

Rekha M. Shangpliang in her chapter "Women and Forest Resource Management: Livelihood Strategies and Interventions" stresses on the fact that during the past five decades of Indian planning, women have been thoroughly marginalized in the process of development, implementation and reorientation. Needless to say, the constraints faced by the women in controlling and managing land and forest resources have to a certain extent deprived them of their claims on sustainable livelihood. The scenario in respect of the Khasi women who claim to belong to matrilineal society is no better either. Though in principle, the matrilineal rule of female inheritance and descent is followed, their customary practices and usages are certainly not gender egalitarian. The overall administration and superintendence of the domestic matters are vested in the mother's brother. The Khasi women inherit property but lack the power to manage it. The woman is the central figure in the household organization, but when it comes to the allocation of the rights over land, a sharp distinction is maintained between 'ownership' and 'control'. While ownership of land is transmitted through women, control invariably lies with men.

Section 3 of the book is titled **Environment and Community Life**, which is comprised of six chapters that reflect on community participation and environment conservation. The first chapter by **Hormila Zhingkhai** on "Enhancing Lives through Environmental Conservation: A Short Narrative of Khonoma Village" focuses on the initiatives taken by the community of Khonoma village to conserve their environment which consequently has led to the enhancement of their social, economic and ecological life. It attempts to look into the challenges and opportunities they faced in their initiative to conserve their environment and bring out the linkage between their conservational effort and their traditional practices and community solidarity which were the main factors behind their success. At the same time, it also tries to examine the various implications of their conservational endeavour.

The joint chapter by **Aabid Hussain Mir** and **Krishna Upadhaya** on "Plant Diversity in Community Managed Forests of Meghalaya" reveals some interesting facts about floristic composition and structure of the forest in Meghalaya. The study was conducted to understand the floristic composition and structure of the forests under traditional management gradient. It also examines the reason for fast disappearance of such forests and suggests effective conservation strategies. The results revealed that the disturbance index was low, species richness, density and basal cover was high in forest involving higher degree of protection than those with lower protection. Such results underline the important contribution of traditional forest management practices of the local people in maintaining a balance between preservation and consumption of forest resources.

Lalmalsawmi Hlondo, in her chapter entiled "Village Forest Development Committee (VFDC) in Forest Management in Champhai Conservancy in Town of Mizoram" is an attempt to make an assessment of the activities of the VFDC in the protection and development of forest lands covered under JFM in Champhai town and thereby understands whether the scheme has successfully been implemented so as to create awareness among the villagers and enhanced environmental sustainability considering the importance of Village Forest Development Committee (VFDC) in bringing about a change in the outlook of the local community and developing the forest lands.

Brinda Bazeley in her chapter entitled "Sustainable Use and Conservation of Forest Resources in Meghalaya" focuses on the sustainability of various forest resources in Meghalaya, and highlights the traditional sacred groves, and certain traditional sustainable agricultural practices in Meghalaya which have been used for time immemorial and has gained attention worldwide.

The joint chapter by **Yodida Bhutia** and **Georgia Liarakou** "Environmental Worldview among the Teacher-Trainees of DIETs, Meghalaya" is based in the District Institute of Education and Training (DIETs), Meghalaya, which are involved in training the prospective teachers for the elementary schools and primary schools. The DIETs are located in each district of the State, which provide training to teachers of primary and elementary schools. Meghalaya has seven DIETs, located in Sohra, Thaladskein, Nongpoh, Nongstoin, Rusubelpara, Tura and Baghmara. The study found that the teacher-trainees have ecological worldview, gender and religion do not affect the environmental worldview. Effect of family, college and internet had least effect on their environmental worldview, however impact of school was highest on forming the worldview of the teacher-trainees. The symbiotic relation between nature and culture is exemplified in many tribal belief systems, art and artifacts which are strongly reflected in their cultural symbols, traits and customs.

Lolita M. Shangpliang in her chapter entitled "The Impact of Border Haats on the Socio-economic Life of the People of Meghalaya" brings out an interesting account of the history of trade relations between Bengal and Surma valley of Assam through Sylhet which existed since Mughal period. During pre-British time transactions were mostly carried on barter system and also through payment by cowrees (sea shells) although some Khasi Syiems are said to have struck coins for internal transaction. Immediately after their arrival the British saw the potentiality of natural resources of Khasi and Jaintia Hills and realized the importance of trade with Bengal. Accordingly they reorganized the markets on a better footing and introduced standard weights and measures in place of prevailing crude barter system. The Government struck coins and converted barter system into monetary one. As the trade further expanded the nature and volume of trade underwent significant changes in the late 18th and early 19th century. The British merchants started lime trade which became a monopoly business for many years to come. Together with lime and coal trade other agricultural produce, oranges, betel nuts, pan leaves produced by the Khasi villagers found easy outlet into

Bengal. Thus the border tribal people in pre-colonial times earned their livelihood by selling agricultural produce most of which being highly perishable like oranges, betel-leaves, vegetables, pineapples etc. Other important agricultural products were bay leaves, pepper, yam, areca nuts etc. The hills are also famous for forest products namely timber, bamboo and lac. Mineral products like lime stone and coal and iron used to be exported to Sylhet in large quantities after the British came to Khasi Hills. Iron smelting and manufacturing agricultural implements and house-hold tools like daggers, knives etc. kodalis was another source of livelihood of the Khasis. These iron tools and implements used to be sold in the border haats and they had good demand in the plains districts. The cottage industry of iron smelting and manufacture of implement had reached a high degree of proficiency even before the coming of the British.

Trade and commerce were important economic activities of the pre-colonial Khasis which were carried out on both internally and externally. They took the inland products to the border markets and exported them outside the State, and at the same time, they bought foreign products and sold them at internal markets located within Khasi Hills. This work makes an attempt to trace the history of border trade between the districts of Sylhet in Bangladesh and Khasi Hills in Meghalaya and to focus on the impact of Balat border Haat, an important border market located in Mawsynram Development Block, on the socio-economic life of the border people of Meghalaya.

The fourth section of the book is titled **Nature, Culture and the Tribes of North-East India.** Altogether nine chapters are covered within this broad theme which range over a wide variety of issues concerning human-nature relationship in the context of North-East India. **Raima Narzary**, in her chapter entitled "Culture of the Karbis in North-East India" focuses on the economy, culture and religion of the Karbis and also on the myths and legends associated with their origin.

Dominic Meyieho in his chapter "An Angami Perception of Nature: A Beckoning for Inter-Cultural Dialogue and Criticism" dwells on the argument that while in the contemporary culture, *the Human* relationship with *the Nature* is that of *subject* and *object*, i.e. *the Human* as mistress/ master and *the Nature* as resource. Contrary to this view, the relationship between *the Human* and *the Nature* from an *Angami* perspective is one in whose culture the differentiation between them is minimal. The Angamis perceive themselves as part of *the Nature* and play with its seasons. To them, *the Divine* is subsumed in *the Nature* and hence, it is not merely "natural", but "sacred". *The Nature* is a living organism and has a soul. She is the sustainer, provider and the way of life. She is not a deceiver – all activities are centred and ordered in and around her. The festivals (*thenyi*) are celebration of *the Nature's* happenings along with *the Human* endeavours in cultivating her for mutual benefits. In the light of contemporary scientific culture and the Angami's perception of *the Nature*, this chapter invites a cross-cultural dialogue and mutual criticism. This is because culture is the locus in which everything is housed. However, there are no cultural universals, but only cultural specifics. Therefore, to secure *the Nature*, dialogue and criticism between cultures is a necessity.

Kiran Barua Nongrum in her chapter "Society and Environment: Nature in the Life of the Khasis" gives a detailed analysis of many natural elements such as streams, rivers, trees, bamboo etc., which are used to procure their utilitarian aspects and are defined in a sacred note this makes the society use nature and its resources with fear and respects that paves a reason for protecting environment. Among the Khasis, the relation between man and environment is rooted in religious context, and this is held collectively to achieve the objective utilitarian function of natural resources on the basis of sacred note. However, today in fast developing, globalizing world and modern setup, Khasi social lifestyle is neutralizing to the changing demands. The growing Western culture and its impact on the new generation have possibly created a gap between the indigenous traditional knowledge and modern way of life. Sophisticated technology, and newly defined gadgets has not only positively benefited the society, but the bottom line in its extreme usage have made our purpose dependent and less innovative and inventive. Tribal diligent behavioural attributes have been swept away from its culture owing to the greater attraction brought by the globalized consumerism lifestyle. Modern globalized action and its trends have engulfed tribal way of life so deeply, which poses a great challenge for traditional knowledge, art and skill which once evenly synthesized meritoriously with nature.

Jawahar Debbarma in his chapter entitled "Man-Nature Relationship of *Borok* Community in Tripura" focus mainly on man-nature relationship of the Tripuri (*Borok*) community through their social, cultural and personal interactions with the nature. It also sheds light on the effect of modern lifestyle on the traditional tribal way of life and its impact. The 'life world' that includes culture and traditions of the tribes are indispensably rooted in the natural environment on which tribes subsist. Tribes of Tripura like any other North-Eastern tribes are influenced by their surroundings and they are nature loving people. Their rituals, belief systems, folk cultures and food habits are extremely definitive of the place and environment. Tripura is the home of a number of ethnic groups which comprised of nineteen sub-tribes, like Debbarma, Reang, Jamatia, Koloi, Mog, Chakma, Lushai etc. Their customs and rituals are more or less the same and they have a similar lifestyle. Though Tripura is a meeting ground of different races, tribals of Tripura reflect distinct traits in their lifestyle. Tribals have a rich cultural heritage of music, fine arts, performing arts, and handicrafts. '*Jhuming*' or shifting cultivation was once not only the prime mode of food production, but also the entire musical, cultural and oral literary traditions revolved around '*Jhuming*'. All the festivities and merry-making were based on '*Jhuming*'. However, with the passage of time people have given up this outdated mode of cultivation and have opted for a stable farming. And distinctly this has an impact on the cultural traditions of the tribals. Some aspects of belief system in the contemporary world may be considered superstitious; however, this belief system indicates a close relationship between man and nature. It is in and through nature that the tribes of Tripura narrate their history, folklore, folk songs and popular proverbs.

Finley Syngai in her chapter "Folklore and Nature in the Context of the Khasis: A Look at Some of the Cave Lores" focuses on the role of folklore in understanding ethno-cultural traits of humans relating to nature. This chapter is focused on the Khasis who have many legends connected with caves.

Sanjeev Shankar in his chapter entitled "Revitalizing Traditional Knowledge in North-East India by Combining Conservation, Research and Entrepreneurship: Living Root Bridge Case Study" stresses on living root bridges or *Ficus elastica* which are based bridges within dense tropical rainforests of Meghalaya in the North-East India. Ranging in span from 15 feet to 250 feet, these bridges are grown by Khasi tribes over a time period of 15 to 30 years, and last for several centuries. With (1) exceptional robustness under extreme climatic conditions, (2) minimal cost, (3) progressive increase in load-bearing capacity, (4) carbon sequestration, (5) remedial properties on surrounding soil, water and air, (6) collective grass root involvement based on human-plant interaction, (7) support to other plant and animal systems, and (8) keystone role of *Ficus* plant species in local ecology, living root bridges offer an extraordinary opportunity for revitalizing traditional living plant based knowledge to ensure long term socio-ecological resilience for tropical regions. Despite these attributes, living root bridges are being replaced by inappropriate solutions owing to increasing resource needs, lack of awareness and the nexus of poverty, population explosion and environmental degradation. The author highlights the value of combining conservation, research and entrepreneurship for revitalizing traditional knowledge, and proposes a socio-scientific-entrepreneurial framework to adapt these bridges as a source of food, medicine and nutrition for local communities.

V. Bijukumar in his chapter entitled "Production of Traditional Ecological Knowledge: Intersecting Culture and Environment" stresses on the interaction between a cultural community and environment. This produces a new form of knowledge called the Traditional Ecological Knowledge (TEK). TEK is often described as the experimental knowledge based on observation and transmitted from one community to another. Although traditional knowledge is an empirical knowledge based on life experiences of human beings with nature, it is often presented out of the sphere of scientific knowledge as it is based on beliefs, traditions and values. While the advent of modernity brought dramatic changes in the relationship between human beings and environment, anthropocentric development created imbalances in the relationship between nature and society, as the development model it puts forward is based on the exploitation of nature for the greed of the people. The onslaught of modernity on environment created an atmosphere of the denial of TEK as irrational and unscientific. Such a disturbing trend poses a grave threat to the sustainability of both nature and human beings. The preservation of TEK is essential for the survival of larger ecosystems and hence the growing realization that it can preserve bio-diversity in a society.

The chapter jointly written by **U. Shilla** and **B.K. Tiwari** entitled "People's Perception on Impacts of Fire and Grazing on the Grassland Ecosystem of Cherrapunjee" is based on a survey, which was conducted in order to obtain the opinion of the local inhabitants about the extent of fire and grazing events and their impacts on the grassland ecosystems of Cherrapunjee. The specific objectives of the study were – to document the extent of fire and grazing in grassland ecosystem; to substantiate the negative impacts of fire and grazing; to develop strategies for prevention of further degradation and eco-restoration of the grassland ecosystems of Cherrapunjee. The result of the study found that fire incidences have been taking place since time immemorial and this phenomenon occurs annually during the dry season. Grazing was mostly carried out in community land which is a common property. When developments of these ecosystems are absent then it can lead to the "tragedy of the commons".

Wazir Alam in his chapter "Perspective on Water Security Status in Manipur: A Comprehensive Review Study" discusses water as a common property resource that has many interlinked dimensions and requires a holistic approach for its conservation and management. The major challenges of water security in present world situation are mainly dependent on its accessibility, afordability and availability of quality water. Although, water is available in various forms on the earth surface, but its spatio-temporal distributions in globe and its interactions with natural systems determine its availability and quality that have further intense interference in determining the quality of life. So, the term 'water security' ensures sustainable water management initiatives that can further ensure security of food, health, society, economy and culture of the people in a region. From last few decades, due to scarcity of both quality and quantity of drinking and irrigation water, the water security is emerging as a major environmental problem in Manipur in spite of the receiving substantial rainfall by the region. The problems of water scarcity, drought like situations and other hydrometerorological disasters such as flood, landslide, droughts, cyclones etc. are very frequently reported from many parts of the state. The chapter is an attempt to review status of water security in Manipur to highlight some of its important issues, challenges and dimensions that requires immediate attention for sustainable management initiatives.

An Overview

The chapter contributors in this volume are predominantly academicians from various disciplines such as Sociology, Anthropology, History, Law, Geography, Creative Arts, and also from the Indian Institute of Management and School of Architecture, which explains the diverse areas that environment can explore. It is interesting to note that the last section on North-East India, which devotes itself to the intrinsic relation between humans and nature and the tribal indigenous faith, expresses concern over the drastic changes taking place in the environment today where issues of 'development', 'environment' nexus has taken centre-stage amongst scholars, academicians and policy-makers. This brings us to the new environmental debate on development options in contemporary India and the ecological limits to growth and development. It is against this backdrop that this seminar was focused on the linkages between culture and ecology. The outcome of the seminar was that it brought together different minds to ponder and think over the issue of sustainability of natural resources. In a country like India where the vast majority of rural households are dependent on the environment to meet their daily household needs, accessibility to natural resources seems to be a crucial problem. The issue of development in North-East India has posed a challenge to planners since Independence. The region predominantly inhabited by tribals has its own opportunities and constraints, but many of the developmental initiatives of the region follow the national perspective of development where the environment was found to be the most potential resource for industrial and infrastructural development. This approach followed by the Government of India has directly affected the livelihood of the rural poor living in the North-East. The tribals of the region depend on bio-mass related products which are mostly available free of cost from the immediate environment and they live within nothing other than a bio-mass based subsistence economy. Another important problem that has surfaced in recent years is the alienation of land from the poor peasants due to privatization of land by a few landowners and acquisition of land by Government. Therefore, a fair and just distribution of land and forest resources has not been entirely successful to construct a development agenda that suits the needs of the rural poor.

It is hoped that the present volume generates new ideas on conservation of environment and initiates useful discussions on the issues raised by the contributors.

Shillong **Rekha M. Shangpliang**

Contents

Foreword by **Prof. David R. Syiemlieh** *v*
Acknowledgements *vii*
Editorial Note – **Rekha M. Shangpliang** *ix*
List of Contributors *xxvii*

SECTION I
ENVIRONMENT AND SOCIETY: SOME THEORETICAL CONSIDERATIONS

1. Environmental Movements in North-East India 3
B. Panda
2. Indigenous People and their Development: Legal and Constitutional Safeguards 16
Chintamoni Rout
3. Ecology, Sustainability and Embedded Belief Systems 25
Sanjib Kakoty

SECTION II
WOMEN AND NATURAL RESOURCE MANAGEMENT STRATEGIES

4. Naga Women as Natural Resource Managers: Is Revival of Tradition Always the Best Option? 33
Lucy Zehol
5. Women and Forest Resource Management: Livelihood Strategies and Interventions 43
Rekha M. Shangpliang

SECTION III
ENVIRONMENT AND COMMUNITY LIFE

6. Enhancing Lives through Environmental Conservation: A Short Narrative of Khonoma Village 59
Hormila Zhingkhai

7. Plant Diversity in Community Managed Forests of Meghalaya 74
Aabid Hussain Mir and *Krishna Upadhaya*

8. Village Forest Development Committee (VFDC) in Forest Management in Champhai Conservancy in Town of Mizoram 90
Lalmalsawmi Hlondo

9. Sustainable Use and Conservation of Forest Resources in Meghalaya 101
Brinda Bazeley

10. Environmental Worldview among the Teacher-Trainees of DIETs, Meghalaya 112
Yodida Bhutia and *Georgia Liarakou*

11. The Impact of Border Haats on the Socio-economic Life of the People of Meghalaya 127
Lolita M. Shangpliang

SECTION IV
NATURE, CULTURE AND THE TRIBES OF NORTH-EAST INDIA

12. Culture of the Karbis in North-East India 137
Raima Narzary

13. An Angami Perception of Nature: A Beckoning for Inter-cultural Dialogue and Criticism 144
Dominic Meyieho

14. Society and the Environment: Nature in the Life of the Khasis 154
Kiran Barua Nongrum

15. Man-Nature Relationship of *Borok* Community in Tripura 171
Jawahar Debbarma

16. Folklore and Nature in the Context of the Khasis: A Look at Some of the Cave Lores 182
Finley Syngai

17. Revitalizing Traditional Knowledge in North-East India by Combining Conservation, Research and Entrepreneurship: Living Root Bridge Case Study 197
Sanjeev Shankar

18. Production of Traditional Ecological Knowledge: Intersecting Culture and Environment 207
V. Bijukumar

19. People's Perception on Impacts of Fire and Grazing on the Grassland Ecosystem of Cherrapunjee 220
U. Shilla and *B.K. Tiwari*

20. Perspective on Water Security Status in Manipur: A Comprehensive Review Study 231
Wazir Alam

Index 244

List of Contributors

Aabid Hussain Mir, Department of Environmental Studies, NEHU, Shillong-793 022, Meghalaya.

B.K. Tiwari, Department of Environmental Studies, NEHU, Shillong, Meghalaya.

Prof. B. Panda, Faculty, Department of Sociology, NEHU, Shillong.

Dr. Brinda Bazeley, Associate Professor, Department of Education, NEHU, Shillong, Meghalaya.

Dr. Chintamoni Rout, Associate Professor & Head, Department of Law, NEHU, Shillong, Meghalaya.
Email: *drchintamanirout @gmail.com*

Prof. David R. Syiemlieh, Former Vice-Chancellor, Rajiv Gandhi University, Former Member and Chairman UPSC.

Dr. Dominic Meyieho, Assistant Professor, Assam Don Bosco University, Guwahati, Assam.
Email: *dominic.meyieho @dbhuniversity.ac.in*

Miss Finley Syngai is teaching Sociology at St. Edmund's College, Shillong.

Prof. Georgia Liarakou, Pedagogical Department of Primary Education, University of the Aegean, Greece.

Hormila Zhingkhai, Research Scholar, Department of Sociology, NEHU, Shillong, Meghalaya.
Email: *jungaa.sen@gmail.com*

Dr. Jawahar Debbarma, Assistant Professor, Department of Philosophy, Ram Krishna Mahavidyalaya, Kailashahar, Tripura. Email: *debbarma.jauhar@gmail.com*

Kiran Barua Nongrum, Head of the Department, Department of Sociology, Umshrypi College.

Krishna Upadhaya, Department of Basic Sciences and Social Sciences, School of Technology, NEHU, Shillong-793 022, Meghalaya.
Email: *upkri@yahoo.com*

Lalmalsawmi Hlondo, Research Scholar, Department of Sociology, NEHU, Shillong, Meghalaya.

Lolita M. Shangpliang is Associate Professor in the Department of Sociology, St. Mary's College, Shillong.

Prof. Lucy Zehol, Department of Anthropology, NEHU, Shillong, Meghalaya.

Dr. Raima Narzary, Assistant Professor, Department of History, Lalit Chandra Bharali, Guwahati (Assam).

Dr. Rekha M. Shangpliang, Associate Professor, Department of Sociology, NEHU, Shillong, Meghalaya.
Email: *rekhakeepintouch @yahoo.com*

Sanjeev Shankar, 4/4 Cooke Town, Wheelers Road Extension Cross, St Thomas PO, Bengaluru-560 084, Karnataka.
Email: *sanjeev @sanjeevshankar.com*

Dr. Sanjib Kakoty, Associate Professor, Indian Institute of Management, Shillong, Meghalaya.

U. Shilla, Department of Botany, Union Christian College, Umiam Khwan, Meghalaya.
Email: *shilla.ublari.5@gmail.com*

Dr. V. Bijukumar, Associate Professor, Centre for Political Studies, School of Social Sciences, Jawaharlal Nehru University, New Delhi-110 067.
Email: *vbijuk@yahoo.co.in*

Wazir Alam, Assistant Professor, Department of Forestry and Environmental Science, Manipur University, Canchipur, Imphal.
Email: *wazir@manipuruniv.ac.in*

Dr. Yodida Bhutia, Assistant Professor, Department of Education, NEHU, Shillong, Meghalaya.

SECTION I

Environment and Society: Some Theoretical Considerations

1

Environmental Movements in North-East India

B. Panda

Introduction

The socio-cultural and economic life-world of human beings is well integrated with their natural environment as they draw economic subsistence from nature as well as use various products of nature in their socio-cultural and religious ceremonies. However, overutilization and exploitation of natural resources create undue pressure on ecology, leading to ecological imbalance and environmental crises. The rapid pace of development, urbanization, industrialization, and new consumerism have brought incessant pressure on the existing natural resources, which resulted in depletion of forest cover, air and water pollution, climate change and the like. Human societies across the globe are indeed experiencing environmental challenges of different kind and magnitude. This, in turn has sharpened the environmental activism. There have been concerted efforts to sensitize and mobilize the masses against the environmental degradation and these mobilizational efforts have often coalesced into mass mobilization and protest movements. Amidst various strategies adopted to fight against environmental crises, environmental movement is being perceived as an effective radical strategy to address the environmental concerns. Through these movements, collective consciousness on ecology is raised which subsequently translated into institutionalization of collective action. The collective action expressed through various protest movements has exerted pressure on the policy-making bodies to devise eco-friendly policies and programmes.

In India, environmental movements of diverse scale and magnitude have erupted. There have been mass mobilization to protest against the construction of dams in river beds (hydro-electric projects), mega developmental and industrial projects besides the resistances to various localized environmental problems and crises. While Chipko, Appiko, Narmada Bachao Andolan are well known environmental movements, there are many relatively lesser known environmental movements, which have taken place in various parts of the country to register the resentment and apprehension of the people. There have been protest movements in North-Eastern region of India too which have raised the issues relating to environment and livelihood. The protest movements against the construction of dams on Tipaimukh and Subansiri are well-known protest movements that the region has witnessed.

Theoretical Perspectives

The environmental crisis could be perceived from diverse perspectives. Perceiving the ensuing environmental problems from the vantage of modernity, it is argued that the process of modernization not only brings rapid transformation in the socio-cultural, economic spheres, but also generates ecological degradations and imbalances. The unabated urbanization, industrialization and neo-liberal developmental model have created incessant pressure on the limited natural resources. The ensuing changes in the society, i.e. the rise in material culture, unequal distribution of resources, widened class disparity, widespread displacement have created conditions for people's resentment. Hence, the ecological crises and socio-economic issues are intertwined which warrant a deeper analysis, critical reflection and a systematic understanding.

It is pertinent in this context to delve into the constructivist approach to perceive the ensuing environmental crisis. The environmental issue is fundamentally a social issue where socio-cultural aspects such as culture, politics and human behaviour carry significant importance. By situating environmental issues within relevant social and political contexts, the advocates of social constructivist approach draw the attention of policy-makers thus, make attempts to contribute towards policy-making on environmental problems (Hannigan, 2006: 32-33). The role of the environmental sociologists for the cause of environment, therefore, assumes significance. Their role become crucial to demonstrate how environmental concerns and problems are related to the dynamic social processes rather than restricting it to and relying on elusive models that merely establish link between the breakdown of ecosystem and social variables (*ibid.*). Portraying environmental crisis as a deep seated crisis,

Beck (cited in Goldblatt, 1996) argues that there is a need for a new level of self-critique and critical reflection. The 'relation of definition', i.e. the rules, institutions and the structure which identify and assess environmental problems and risks (*ibid.*, 166) "must undergo epistemological and democratic revolution" (*ibid.*, 187). The ecological concern and environmental problems affect the entire humanity and hence, corrective measures need to be devised and concerted efforts should be made to create awareness across the groups and communities. Therefore, collective action and collective orientation towards this very problem is presupposed. Collective consciousness remains a powerful social force to reckon with through which collective identity and collective actions are articulated, and in the context of ecological crises, the collective action and collective articulation of the problems become all the more important. When ecology is reflected in ideology and consciousness of the people, it creates cognitive and motivational drive which makes people ecologically more conscious and assertive and these subsequently are manifested in their behaviour and practices. This becomes more pronounced in the context of modernity. Eder (1996: 204-206) articulates that ecological consciousness of the people becomes a powerful ideological tool particularly in the context of modernity due to overemphasis on material values which directly affect the environmental ethics and 'environmentalism' (environmental consciousness). This in turn takes a new ideological twist wherein "environmentalism becomes a new ideological tradition in addition to the ideological cleavages of advanced modern societies". Ecology therefore, has become a master frame, a symbolic package in these societies (*ibid.*). This could be seen as a dominant entity that helps forming and perpetuating identity as people tend to demonstrate their growing concerns over ecological crises and subsequently relate these to their very survival and livelihood issues. This is reflected more in today's world; in the attitudes of people as they increasingly become assertive to protect their environment and stake claim over the resources such as land, water and forests which get eroded primarily due to increasing emphasis on materialistic mode of development. Therefore, land, forest and resource use become binding elements as people eke out their livelihood and the resources in nature keep providing sustenance for them. Pointing the nature-culture dynamics in the context of Khasi society Nongkynrih (2012) articulates, one may observe that environment consists of three parts such as mother earth and nature, celestial subjects and human beings in Khasi world-view. Presenting the nature-culture relations of Khasis, Nongkynrih (2012) explains how nature is well integrated with socio-religious life of the Khasis. For Khasi, the environment is referred to as *Ka Mei-ramew* (mother earth) where *Ka Mariang* (nature or earth), celestial entities such as *Ka Sngi* (sun), *U Bnai* (moon) and *Ki Khlur* (stars) and *Ki Khun Bynriew* (human inhabitants of mother earth) are integral parts. In Khasi world-view, *Ka Mariang* constitutes elements such as *Ki Khlaw Ki Btap* (forest lands), *Ki Khyndew-Ki Shyiap* (lands), *Ki Lum-Ki Wah* (hills, peaks, rivers, etc.), *Ka Lyer* (air), *Ka Por-Ka Samoi* (seasons) which are paramount (*ibid.*, 45).

The symbiotic relationship that exists between nature and human beings subsequently becomes a cementing force for the identity assertion. Ecology and the prevailing ecological problems create the conditions for establishing a 'global identity' by transcending geographical and social boundaries. It remains the only global identity on behalf of all human beings, regardless of their specific attachment to social, historical, gender or their religious faith (Castells, 2010: 185). Therefore, to examine and analyse the ensuing environmental crises, environmental activism and environmental movements, the notion of collective action and new social movement perspective assume significance.

Collective Action

Collective action has been an important analytic concept in classical sociological constructs and theoretical formulations. Emile Durkheim's collective conscience, Karl Marx's class consciousness, Max Weber's *verstehen*, Ferdinand Tonnies's *gemeinschaft* provide conceptual frameworks to examine and analyse the notion of collective social action. Collective members were believed to internalize qualities of groups they belong thus creating a unified collective experience, a single canvass against which social actors construct a sense of self (Cerulo, 1997). Avino (2006: 324) points out three aspects which are significant in this context. These are: (i) the transformation of traditional socio-economic organizations (the emergence of networks), (ii) the territory (national become global), (iii) power (the definition of politics and nation-states). This activism as Avino articulates leads to transformative processes wherein constant interaction takes place among 'interacting structures'. These interacting structures include the "different levels and changing contexts of mobilized actions; the activist's interpersonal relationships; the construction of a collective identity; the interaction among organizations and groups; the interaction among social movements and political agents (Avino, 2006: 330). These structures create the suitable platforms for social mobilization and political transformation. One can observe a "widespread surge of powerful expressions of collective identity that challenge globalization and cosmopolitanism on behalf of cultural singularity and people's control over their lives and environment" during the last three decades or so (Castells, 1997 cited in Jayaram, 2012: 44).

Environmental Movement as New Social Movement

New Social Movements (NSMs) tend to target systemic reforms and seek transformative changes through diverse mechanisms and involvement of multiple actors. Various mechanisms such as creating awareness, mobilization of resources and masses, creating a strong chain of vertical as well as horizontal networks, involvement of multiple actors through which NSMs seek transformatory changes. NSMs project new identities and lifestyles (Polleta and Jasper, 2001) to seek changes in the prevailing socio-cultural systems. Environmentalism and other NSMs have claimed to transcend old classical divisions even if activists themselves tended to be middle class or highly educated (Touraine, 1981 cited in *ibid.*). Environmental movements that aim to save the environment from rapid degradation are being seen as an effective strategy to create environmental consciousness among the people which has emerged as an important political force and a source of identity (*ibid.*).

Perceiving environment movement as a NSM, Habermas (cited in Goldblatt, 1996: 126) stresses on three important contemporary issues to examine ensuing environmental crisis. These are: (i) the problem of excessive complexity; (ii) the tangible material destruction of environment that leads to environment politics; and (iii) overburdening of the communicative infrastructure. Environmental movements emerging as an inevitable social action can be attributed to the ecological consciousness emerging from the changing societal condition, lack of comprehensive understanding of the world and the bureaucratization of the will-forming process in the public sphere. Ecological movements in a way sensitize us to the intricate relation that exists between nature and human beings which would not have been explored without these (Giddens, 1987 cited in *ibid.*). Habermas in his theory of communicative action argues that it is not merely the material reproduction but the systematic complexity along with the increasing capacity of a society to respond to its environment are key to develop a critical understanding on ecology (Goldblatt, 1996:117). As environmental movements strive for reforms in the prevailing practices and seek transformative changes multiple actors that include students, intellectuals, and activists along with affected population play crucial role towards the success of these movements.

The success of a social movement depends on the effective modes of communication, mobilization and institutionalization. It begins with communication and networking, consolidates through mobilization of resources and culminates through institutionalization of collective action. The mode of communication for environmental movements heavily depends on traditional networks such as village community, caste, tribe, and lineage (Guha, 2000: 106). These movements are facilitated by community based grassroots organizations, existing intermediary institutions and civil society organizations that promote people's participation, people's mobilization and people's empowerment. Through their intense networks, persistent campaigns and mobilizational efforts civil society organizations are seen as facilitators of NSMs. Finger's (1994: 64) "learning our way out" formula emphasizes on traditional problem solving strategies where Non-Governmental Organizations (NGOs) tend to play a crucial role in this process in order to overcome diverse crises that the society confronts with. Environmental crises can also be effectively addressed through this process. Hence, collective learning units must be promoted in the society where knowledgeable as well as general public would work together and devise appropriate strategies. NGOs could be agents of this social learning mechanism and contribute significantly to this endeavour.

Environmental Movements in India: Definitional and Conceptual Aspects

Environmental movement can be perceived as an organized collective social action which is aimed at creating awareness on environmental problems, fighting against environmental degradation and advocating for sustainable use of natural resources. While on the one hand, these movements raise micro action plans to safeguard natural processes and the environment on the other hand offer macro concept of ecological development at the global, rational and regional level (Bondyopadhyaya and Shiva, 1988: 1232). In India, ecology movements can be seen as the people's response to the threat to their livelihood and survival and "as a demand for the ecological conservation of the vital life support system" (*ibid.*, 1224). "People's movement for clean air and water is growing in all corners in the country just as ecologically irresponsible resource hunger of the process of growth is moving deeper in the hinterland in search of newer resources" (*ibid.*, 1225). Hence, the pressure exerted on limited natural resources due to insatiable developmental aspirations and subsequent exploitation of natural resources have resulted in new political struggle for safeguarding the interest of the poor and the marginalized sections of the society that include women, tribals and the poor peasants (*ibid.*, 1988: 1224). Therefore, ecology movements could be seen as people's response to their very survival. Poverty and under-development, and the issues that emerge out of the very process of development are primarily responsible for environmental movements (Mathew, 1982: 239). Pointing out the diverse ideological strands on ecology, Guha (1988) articulated three ideological strands visible in environmental movements in India. These are: Crusading Gandhian, Appropriate Technology and Ecological Marxists. Crusading Gandhians emphasize community development strategy where traditional village community is seen as a prime protector of environmental and social harmony. Gandhian perspective rests on village development which it is argued, is environment sensitive and effectively addresses environmental problems. Appropriate Technology strand is the synthesis of industry and agriculture where combination of modern and traditional technology ensures large and small units of production. Marxian perspective rests on the premise that environmental degradation is rooted in deep seated inequalities. Therefore, systemic economic change is necessary prior to the ecological stability. Making a distinction between political economy and moral economy Baviskar (1995: 42) articulates that environmental movements gather strength and momentum from these two domains as the environmental movements in India clustered around these two forms of economy, i.e. political economy for profit and moral economy of need. Ecology movements therefore, could be perceived as reaction against the deteriorating socio-ecological condition resulting due to ensuing development process. Putting ecological movements in a perspective Bondyopadhyaya and Siva argued that the classical paradigm of development; resource use created by politics, economics, science and technology are to a great extent responsible for the condition that leads to ecology movements rather than particular development projects (Bondyopadhyaya and Siva, 1988: 1225). They further argued that while commercial forestry generates revenue, it also reduces the multipurpose biomass productivity and damage the hydrology of the forests. People dependent on non-timber biomass of the forests such as leaves, twigs, fruits, nuts, medicines, fuels, oils etc. find it exploitative and extremely difficult to sustain themselves due to commercial exploitation of forests (*ibid.*, 1224). Commenting on the nature and dynamics of the conflicts over natural resources Gadgil and Guha classified the population into three categories, i.e. omnivores, ecosystem people and ecological refugees. Omnivores are those who tend to capture, transform and use natural resources from a wider catchment area and take advantage of their social power. Ecosystem people depend heavily on natural resources of their locality for survival and Ecological refugees live on the margins of prosperity who neither afford to lead a comfortable material life nor could freely pick up from nature (Gadgil and Guha, 2000:4-5). There indeed exists a complex dynamics among these diverse groups with diverse interests. In India, the resistance from ecosystem people could be perceived as the environmentalism of poor (Guha, 2000). Therefore, one can observe that the environmental issues are intertwined with the issues relating to livelihood and poverty of the people.

Trajectory of Environmental Movements in India: A Brief Sketch

Environmental movements in India revolve around competing claims over renewable natural resources and the struggle for the rights of victims due to environmental destruction. These could be seen as a form of new, organized protests in India, which have grown in frequency and intensity that involve people from all social strata (Swain, 1997: 819). The popular environmental movements such as Chipko, Narmada Bachao Andolan have drawn the attention of development planners, environmentalists, civil society organizations, media and the common citizens. Most of these environmental movements while on the one hand have addressed the environmental concerns they also have raised the issue of livelihood of the affected population. Most of these movements have erupted due to development induced displacement which also severely affects the local eco-system. Apart from these well-known environmental movements there have been protest movements against the destruction of forests and to protect the traditional livelihood of local communities in the tribal belts, particularly in central India. Environmental movements in India have also been directed against construction of mega dams on the Kosi, Gandak and Tungabhadra rivers and the canal irrigated areas of Punjab and Haryana (Karan, 1994: 33). As the river valley projects tend to submerge the agricultural land, affect the flora and fauna and cause large scale displacement of population, the environmental protest movements have gained momentum. Besides displacement, issues such as deforestation, waterlogging, downstream water shortages, siltation, and salinization motivate the activists and protest groups to form as a cohesive force (Swain, 1997: 819-825). The establishment of industrial and commercial farms have come at the cost of deforestation causing extinction of many endangered species and medicinal plants. From locally based opposition to individual dam projects, these movements have moved to a more general critique of the state's path toward development, and they are now exploring the viability of 'alternative' modes of development (*ibid.*, 828). Involvement of activists and voluntary organizations provide added momentum to these movements. Therefore, these environmental movements could be seen as the reactions against the narrowly conceived development which demands intensive use of resources which have led to ecological destruction and economic deprivation (Karan, 1994: 33).

The Environmental Movements in North-East

North-East India is an ecologically rich region known for its mountains and hills, water bodies, dense and green cover with diverse flora and fauna. The region has one of the richest and varied flora and fauna in the world (Myers *et al.*, cited in Sinha, 2012: 164-165). The region is recognized as part of the Indo-Myanmar bio-diversity hotspot, one of 25 such hotspots in the world. About 8,000 species of the plants of India are found in this region. Out of the 315 families reported from the entire country, 200 families are found in this region. Arunachal Pradesh is the richest state with regard to the distribution pattern of species with about 5,000 species of flowering plants followed by Assam and Meghalaya (Bhattacharjee, 2010 cited in Sinha, 2012: 165). Bestowed with rich natural resources, the people of this region believe to enjoy very special symbiotic relations with nature where nature plays a very significant role in shaping the socio-cultural worldview of various tribal communities in the region. The socio-cultural life of the tribal communities in the region is therefore, integrated with the nature and natural surroundings where religion, festivals and art linked to nature in one way or the other. Presenting the ecological heritage of Meghalaya, Marwie (2001: 16-17) observes that hills in the state of Meghalaya are home to a wide variety of plants and flowers. One of the biggest collections of orchids is found on these hills. There are as many as 325 species of orchids besides precious medicinal plants and herbs found in abundance, which shows its ecological significance. North-Eastern part of India which is considered as the hub of biodiversity due to its topography and climatic condition has also faced environmental problems which have caused concerns. As Shangpliang (2010: 69) in the context of Khasi society articulates, there has been remarkable change in the pattern of economic pursuit of the Khasis over the past few decades. The forest-based industries such as saw mills, carpentry workshops, plywood and veneer factories have mushroomed in Meghalaya which resulted in exploitation of forest resources.

Growing pace of industrialization, urbanization, commercialization and deforestation are increasingly creating pressure on the region's pristine ecology. Destruction of the tropical rain forests, shrinking biodiversity and wetlands, soil erosion with landslide have taken toll on its ecological system. The region has indeed witnessed various environmental movements to preserve the environment with diverse intensity. Even during colonial rule Garo Hills witnessed environmental movements under the leadership of Sonaram Sangma against the colonial policy which debarred the local tribal communities from jhumming, hunting and collecting forest products. This policy which was exploitative in nature motivated the people to become assertive and raise voice against it. Though the movement was subsequently suppressed, there were occasions where protests withstood this repressive method (Sinha, 2012). Of late, the region has also witnessed resistance movements over river valley projects. Dams constructed on the river Brahmaputra, Barak, Loktak, Tipaimukh, Gomti have led to resistance and protest movements in the region. Construction of dams on Barak valley in Assam, projects on river Teesta and her tributaries in Sikkim, Subansiri river valley project and Siang river project in Arunachal Pradesh (Sinha, 2012) have triggered for the mobilization coalescing into movements. Hence, displacement and environmental degradation caused by dams is considered to be the major cause for environmental protest movements in the region.

The environmental movements in this region however should be perceived differently due to its multi-ethnic set-up and socio-cultural diversity. Different ethnic communities conscious of their ethnic identity resort to diverse strategies to protect their distinct socio-cultural identity and safeguard their interests. Despite various protest movements that the region has witnessed, it is argued that there is absence of high intensity environmental movements which may be attributed to the widespread perception that the region is environmentally not disturbed much and the hilly terrains are covered with dense forests. The lack of effective organizational set-up, mobilizational efforts and other predicative factors such as the practice of jhumming for cultivation, commercial timber for its economic development could also be attributed to this (Sinha, 2012: 177). Further, ethnic movements, prevailing fragile law and order situation could be other factors which might have overshadowed ecology related protests. Commenting on the ambivalences with regard to protest movements against the exploitation of forests in the region, particularly in the context of Meghalaya, Sinha (2012) asserted that the newly emerged elites are largely responsible for the depletion of forests. Besides, absence of effective forums to fight for the protection of forests despite forests playing a significant role in the lives of the tribal communities is baffling. In spite of depleting forest cover, there are many noticeable awareness programmes and grassroot movements to safeguard forests thereby to strive towards establishing a healthy environment (Sinha, 2012: 191). For the tribal neo-elites in the wake of modernization (urbanization and industrialization, material culture), the forest is merely turning into a storehouse of trees providing commercial timber for marketing (*ibid.*). Therefore, one may not observe environmental movements with the same intensity and dimension as it has happened elsewhere. While we delve into the analysis of environmental movements in the region, the distinct socio-cultural milieu, pace of growth and development, rich biodiversity are some of the aspects which need to be kept in mind.

Conclusion

The deep socio-cultural and economic bond that exists between nature and human beings suggest that human beings are the problem as well as solution for the environmental problems. It is they who have to devise appropriate strategies to restrain their insatiable greed emanating from modernity and ensuing socio-economic processes of change such as industrialization, urbanization, population growth, material culture and consumerism in the society. Therefore, there is need for concerted and continued efforts from diverse stakeholders to save the environment from its degradation. Among various strategies, environmental movement could be seen as a radical strategy towards this direction. Environmental movements that have taken place in various parts of India demonstrated that environment and livelihood are intrinsically connected and the healthy environment is instrumental for healthy living. These mobilizational efforts which often have culminated into movements have been able to wedge a sustained pressure on the policy-makers to carefully weigh the ecological concerns and plight of the affected and displaced population before executing mega developmental projects. These movements have also created a sense of ownership among the affected population and made them assertive to protect their land, hearth and resources. The environmental activism witnessed in North-East India has also reflected the environmental consciousness of the people in the region.

REFERENCES

Avino, E.D.P. (2006). "Social Movement in a Globalized World", in Samir Dasgupta and Ray Kiely (eds.), *Globalization and After*, Sage Publications, New Delhi, pp. 320-48.

Baviskar, A. (1995). *In the Belly of the River: Tribal Conflicts over Development in the Narmada Valley*, Oxford University Press, New Delhi.

Bondyopadhyaya and Siva (1988). "Political Economy of Ecology Movements", *Economic and Political Weekly*, June, 11: 1223-1232.

Castells, M. (2010). *The Power of Identity*, Wiley-Blackwell, UK.

Cerulo, K.A. (1997). "Identity Construction, New Issues, New Directions", *Annual Review of Sociology*, Vol. 23, pp. 385-409.

Eder (1996). "The Institutionalisation of Environmentalism: Ecological Discourse and the Second Transformation of the Public Sphere", in Lash *et al.* (eds.), *Risk Environment and Modernity: Towards a New Ecology*, Sage Publications, London.

Finger, M. (1994). "NGOs and Transformation: Beyond Social Movement Theory", in Princen, T. and Finger, M. (eds.), *Environmental NGOs in World Politics: Linking the Local and the Global*, Routledge Publications, London, pp. 48-66.

Gadgil, M. and Guha, R. (2000). *The Use and Abuse of Nature*, Oxford University Press, New Delhi.

Goldblatt, D. (1996). *Social Theory and the Environment*, Polity Press, UK.

Guha, R. (1988). "Ideological Trends in Indian Environmentalism", *Economic and Political Weekly*, Vol. 23, No. 49, pp. 2578-2581.

Guha, R. (2000). *Environmentalism: A Global History*, Oxford University Press, New Delhi.

Hannigan, J. (2006). *Environmental Sociology: A Social Constructionist Perspective*, Routledge Publications, London.

Jayaram, N. (2012). "Identity, Community and Politics: A Survey of Issues and Analysis", *Economic and Political Weekly*, Vol. XLVII, No. 38, pp. 44-61.

Karan, P.P. (1994). "Environmental Movements in India", *Geographical Review*, Vol. 84, No. 1, pp. 32-41.

Mathew, T. (1982). "Governmental Response to Environmental Needs in India", *India International Centre Quarterly*, Vol. 9, No. 3/4.

Mawrie, B.L. (2001). *The Khasis and their Natural Environment: A Study of the Eco-consciousness and Eco-spirituality of the Khasis*, Vendrame Institute Publications, Shillong.

Nongkynrih, A.K. (2012). "Tribes and Their Environment: A Sociological Interpretation", *Oriens Journal*, Vol. III, pp. 42-60.

Polleta, and Jasper, (2001). "Collective Identity and Social Movements", *Annual Review of Sociology*, Vol. 27, pp. 283-305.

Shangpliang, R.M. (2010). *Forest in the Life of the Khasis*, Concept Publishing Company (P) Ltd., New Delhi.

Sinha, A.C. (2012). *Colonial Legacy and Environmental Crisis in North-East India*, Eastern Book House, Guwahati.

Swain, A. (1997). "Democratic Consolidation? Environmental Movements in India", *Asian Survey*, Vol. 37, No. 9, pp. 818-832.

Tourine, A. (1985). "An Introduction to Social Movements", *Social Research*, Vol. 52, No. 4, pp. 749-787.

2

Indigenous People and their Development: Legal and Constitutional Safeguards

Chintamoni Rout

Introduction

The Universal Declaration of Human Rights was adopted by the UN General Assembly on 10th December, 1948. Since then long time has elapsed, but absolute protection of all the human rights have not been achieved till now. Particularly, the human rights of the disadvantaged or vulnerable groups, like women, children, disabled, *dalits*, aged, minors, Scheduled Castes and Scheduled Tribes need special protection in the international and national fields. Human rights are the Fundamental and Inalienable Rights. The preamble to the Universal Declaration of Human Rights, 1948, emphasizes that, "recognition of inherent dignity and of the equal and inalienable rights of all members of the human family is the foundation of the freedom, justice and peace in the world". The concept of human rights is closely related with ethics and morality. They are also termed as Universal moral rights.

Over the past 70 years, the International Movement of Indigenous People has grown with extraordinary speed. Today, indigenous peoples have emerged as a distinct category in international human rights. Indigenous peoples around the world have sought recognition of their identities, their ways of life and their right to traditional lands and resources. In September, 2007, the General Assembly of the United Nations has adopted "Declaration on the Rights of Indigenous Peoples". Article 1 of the Declaration declares that all "indigenous peoples have the right to the full enjoyment, as a collective or as individual, of all human rights and fundamental freedoms as recognized in the Charter of the United Nations.

Thus, the struggle of the world's indigenous people has reached a new height and the responsibility is upon the Nation-States to respect the rights of the indigenous people to maintain their identity, culture and interest to provide an environment to participate in social, economic and political life.

Since India became independent and the Constitution of India declared it as a 'welfare state', it became the primary duty of our National Planning Commission to ensure the development of the poor masses suffering from hunger, poverty, disease and illiteracy. Special provisions were included in the constitution to safeguard the social, economic, educational and political interests of weaker sections of the community including tribals. The people of tribal community had been treated as those living far away from civilization in forests and hills. But, after Independence, the policy of *laissez-faire* followed by the British could no more be continued. The tribals also had the right to come in line with society and for that the gap between them and rest of the people had to be bridged.

North-East Indian Tribal Population

North-East India is well-known for its composition of population and cultural pluralism. This region accommodates large number of tribes. There are hill tribes, plain tribes and the people of the plains. Each group of population is heterogeneous in character. The culture and the way of life of the hill people are totally different from those of the plain areas. The diversity in culture, language and religion was not active towards the destruction of social cohesion for a long time. Before recognition of this region, i.e. political separation, Nagaland, Meghalaya, Mizoram and Arunachal Pradesh were hill districts of the State of Assam. Since 1960s, the process of reorganization of this region was started and accordingly, Arunachal Pradesh, Mizoram, Meghalaya and Nagaland have been carved out of the State of Assam. Even after political separation these states are treated as one region for the developmental purpose and other considerations. The interdependence of the North-Eastern States cannot be ignored and the State of Assam is the gateway to the region. Considering this factor of interdependence of the states, the Government of India established North-Eastern Council immediately after reorganization of this region and it is treating it as one region for developmental programmes. In spite of distinct socio-cultural differences and political separation these states have several common problems and requirements.

Another significant point is that except the hill areas of Assam like Karbi Anglong and North Cachar Hills districts, all the tribal areas are located on the international border lines. For example, Kameng, Subansiri and Siang districts in Arunachal Pradesh have a common boundary with Myanmar. Mizoram has a common boundary with Bangladesh and Nagaland has a common boundary with Myanmar. All these states before becoming the full-fledged states were the parts of the State of Assam. Therefore, one finds that undivided Assam had common international boundary at several areas, which were inhabited by the hill tribes with different culture, language and religion.

Again diversity is noticed in indigenous political institutions. These may be grouped into main three categories – authoritarian, republican, and semi-monarchical. The tribes may be divided and sub-divided into different groups. Further there are some small groups of tribes which have their own culture and way of life. Apart from racial diversity, there is also linguistic diversity. There are several hundreds of language and dialects spoken by different tribes in this region. One may view this kind of pluralism as resource to be used for the progress of the people of this region which may be preserved for the growth of human civilization.

India's tribal population is almost equal to or even more than the total population of the United Kingdom. It has the second largest tribal population in the world, next to Africa. There are 630 tribes, sub-divided into groups, spread over the length and breadth of the country, concentrated mostly in the hilly and forest regions. The tribes differ considerably from one another in race, language, culture, customs and traditions, in their myths and beliefs, and present a spectacle of striking diversity. This diversity, marked by varied racial characteristics and diverse cultural traditions and linguistic traits lend lustre to the cultural mosaic of our country and they represent the true synthesis of India's culture.

In spite of much diversity, there are certain broad similarities between the mutually divergent groups. Striking similarities are noticed in their way of habitation, mode of living, social customs of marriage and divorce, religious beliefs, performance of festivals. It ordinarily has a leader, an ancestor as well as village deity. A tribal society is usually small, isolated, closely knit and is largely organized on the basis of kinship. The most important common feature is their simplicity, honesty and truthfulness.

Looking at the cultural traditions and beliefs of tribals, the problem that has been exercising in the minds of thinking persons in India, especially after attainment of independence, is what should be place of tribal people in the framework of Indian nation and how they should be developed and brought to a level with the rest of the people – socially, economically, culturally and politically. Before independence, tribals were exploited by landlords, forest contractors, liquor venders, revenue, excise and police officials, robbed by moneylenders, cheated by merchants and their culture was largely destroyed by Christian missionaries. The tribal communities were left in isolation by the British who followed the policy of *laissez faire* in respect of tribals. Complete isolation has never led to progress and advancement, but always to stagnation. The British policy transformed the tribal's relationship with the forest. Tribals had depended on forests for food, fuel and cattle feed. They practiced shifting cultivation, known as *jhum*, *pudu* etc. The tribals also thought they were better-off being isolated and leading a "happy go lucky life". They were away from the so-called 'civilized society'.

Strategy for Tribal Development

In most tribal areas, resources are plentiful and the economic needs of the community are comparatively limited. Tribal social and economic-organization is relatively simple. They have a strong tradition of self-governance; managing the day-to-day affairs of the community with considerable equity and dignity and without external supervision and interference.

Economic planning in India has stressed on fulfilling the basic needs of the tribal population and thereby ensuring social and economic welfare of these communities. With such objective in view, Tribal Development Blocks were started in 1956 with special emphasis on individual beneficiary schemes. The Commissioner for Scheduled Areas and Scheduled Tribes had commented in his report in 1961 that "food for the year, drinking water during summer, employment for time to time, fuel, wood for repair of house and facilities for minor irrigation are the main needs of tribals". The success of tribal development in the first phase depended mainly on grants and incentives. Development programmes were not in accordance with geographical context and requirements of tribals.

In the second phase of tribal development, an Area Development Approach was adopted during the Fourth Five Year Plan. Effort was made for economic development of tribals with emphasis on agriculture and animal husbandry. During the Fourth Five Year Plan, need for development of cottage and small industries was also recognized for diversification of the economy. Incentives and grants were granted for capital goods and working capital. Due to illiteracy and lack of skills, simple tribal families could hardly benefit out of such policy. Those connected with the mainstream affairs and the non-tribals took advantage of this. Consequently, no specific change was noticeable in the economic development of tribal families. The Fifth Five Year Plan was the third phase of tribal development when the concept of tribal sub-plan was adopted for the first time. To speed up development of tribal areas, the special provisions like, a definite proportion of total proposed state plan was earmarked for Scheduled Area and integrated tribal development projects were to be formulated for sub-plan area. The integrated tribal development projects were prepared by combining special central assistance, state plan funds, and institutional finance at DRDA level for implementation so that local needs could be satisfied and local resources could be fully exploited. People's representatives were also consulted at local level while framing schemes at local level.

The office of Secretary and Commissioner, Tribal Area Development was created in 1975-76 so that special attention is given to problems of tribal welfare and advancement. The Modified Area Development Approach (MADA) was adopted in 1978-79 by identifying MADA clusters of population of 10,000 having more than 50 per cent tribal population. Subsequently, in 1986-87, pockets of tribal concentration of 5,000 population were also identified for development of scattered tribals residing outside Tribal Sub-Plan (TSP) area. The Commissioner, TAD identified the needs of tribals and prepared proposals for schemes and programmes to be included in TSP every year. Accordingly, area development projects are undertaken and the programme includes various aspects like; agricultural development, horticulture development, plant protection, goat development, cattle development camps, sheep development, minor irrigation, electrification of tribal localities, education, medical and health and also economic assistance. The focus was on raising the economic status and fuller realization of the development potential of tribals and the tribal areas with the objective of giving them social and economic justice.

Constitutional Safeguards

The Constitution of India also provides for a number of special provisions for the Scheduled Castes and Scheduled Tribes. As per the provisions contained in Article 342, a schedule of tribes with reference to each state/union territory is notified by the president. Article 244 provides for special provisions for Scheduled Areas included in the Fifth and Sixth Schedules. Ten states like, Andhra Pradesh, Bihar, Jharkhand, Himachal Pradesh, Gujarat, Maharashtra, Madhya Pradesh, Chhattisgarh, Odisha and Rajasthan are included in the Fifth Schedule and the so-called seven sisters of Assam, Arunachal Pradesh, Nagaland, Manipur, Mizoram, Tripura and Meghalaya in the North-East are covered by the Sixth Schedule. Article 17 makes two declarations. First, it says that "Untouchability" is abolished and its practice in any form is forbidden. Second, enforcement of any disability arising out of "Untouchability" has been made an offence punishable in accordance with law made by Parliament. Parliament enacted the Untouchability (Offences) Act, 1955, prescribing punishment for practising untouchability. This Act was renamed in 1976 as the Protection of Civil Rights Act, 1955 and given more teeth.

The Supreme Court of India in Kailash Sonkar *vs.* Smt. Maya Devi AIR 1984 SC 600, and N. Adithayan *vs.* Travancore Devaswom Board, AIR 2002 SC 3538, taking support from *Shrimad Bhagavad Gita* and Mahatma Gandhi's views has observed that all distortions based on caste and creed must be abolished and man must be known and recognized by his actions irrespective of the caste to which he may on account of his birth belong. The Court referred to Article 17 reflecting the vision of the founding fathers of the constitution to liberate the society from blind and ritualistic adherence to more traditional superstitious beliefs. The evil of untouchability has gradually been slackened with educational and economic upgradation of the oppressed classes. To root out untouchability and its varied forms, the Parliament has enacted the Scheduled Caste and Scheduled Tribes (Prevention of Atrocities) Act, 1989. Article 23 prohibits traffic in human beings and beggar, and similar other forms of forced labour. In pursuance of it, the Bonded Labour System (Abolition) Act, 1976 was enacted. Though the Act is general, it prevents exploitation of weaker sections including Scheduled Castes and Scheduled Tribes. Article 24 prohibits child labour below 14 years of age and prevents the exploitation of Scheduled Castes and Scheduled Tribes whose children are engaged in hazardous occupations, being poor.

Educational and Cultural Safeguards

Article 15(4) empowers the state to make special provisions for the advancement of any socially and educationally backward classes of citizens including Scheduled Castes and Scheduled Tribes. There are yet another vulnerable groups of citizens which because of their vulnerability or otherwise need special treatment. Clause (4) of Article 16 is another exception to Clauses (1) and (2) of Article 16. Clause (4) expressly makes provision for reservation in state services. It states that the state may make reservation of appointments or posts in favour of any 'Backward Class' of citizens, which, in the opinion of the state is not adequately represented in the public services under the state. Article 335 provides for reservation of posts for Scheduled Castes and Scheduled Tribes considering maintenance of efficiency in administration in making appointments to service and posts in connection with the affairs of the Union and the States. Article 350(A) provides for adequate facilities for instructions in mother tongue at the primary stage of education to children belonging to linguistic minority groups.

Political Safeguards

Article 332 in Part VI of the Constitution provides for reservation of Seats for Scheduled Castes and the Scheduled Tribes in the Legislative Assemblies as nearly as possible to the proportion of their population in the state. Article 335 has made provision for the claim of members of Scheduled Castes and Scheduled Tribes consistently with the maintenance of efficiency of administration, in making of appointments to services and posts in connection with the affairs of the Union and the States. Article 338 makes mention of special office for protection of their interests known as the National Commission for Scheduled Castes and Scheduled Tribes. The functions of the Commission include investigation, monitoring and evaluation of various safeguards provided to Scheduled Castes and Scheduled Tribes by the Constitution and other Acts passed by the Union and the States.

Economic Safeguards

Article 244 of the Constitution of India contains provisions as to the administration and control of Scheduled Areas and Scheduled Tribes. Part A of Para (3) provides that "the Governor of each state having

Scheduled Areas therein shall annually, or whenever required by the President, make a report to the President regarding the administration of the Scheduled Areas in that State and the executive power of the Union shall extend to the giving of directions to the state as to the administration of the said areas". The Fifth Schedule contains provisions regarding the administration and control of Scheduled Areas and Scheduled Tribes. Article 275(1) also provides that "there shall be paid out of Consolidated Fund of India as Grants-in-Aid of the revenues of state such capital and recurring sum as may be necessary to enable that state to meet the cost of such schemes of development as may be undertaken by the state with the approval of Government of India for the purpose of promoting the welfare of Scheduled Tribes in that state".

Conclusion

Thus, after Independence, since India accepted development approach as part of its ideology of a welfare state, infrastructure development works were undertaken on a large scale in tribal areas. Many problems concerning tribal development like displacement, poverty alleviation, health and disease, land alienation, indebtedness, criminalization etc., are in fact the resultant effects of development. It is known that due to the large irrigation and power projects there is displacement of poor tribals, rendering them landless and unemployed, there are disparities between rich and poor, nature and civilization. Movements launched by social activists like Sunderlal Bahuguna in Garhwal, Baba Amte and Medha Patkar in Madhya Pradesh and Maharashtra are quite well-known. The modernization model of development has disturbed the 'man-nature-society' relationship in tribal areas.

The Land Acquisition Act, 1894 provides for such compensation. Rehabilitation policy for oustees was not well formulated. Many of the mainstream projects, while bringing development for the non-tribals, brought misery for the tribals. Prior to 1854, forests were not scarce commodity and tribals were traditionally enjoying forest rights. By enactment of the Forest Act, 1864, the government took away all the customary rights of tribals. They were allowed to cultivate forest lands only by paying fines. But, now, with the passing of the Forest Conservation Act, 1980, tribal have become intruders and even their agreed possession prior to July 1, 1980, are still unregularised. There was a persistent need to associate the beneficiaries of programmes in formulation, execution and monitoring plans prepared for their development.

Tribals are part of the Indian society, but at the same time they are different. Hence, any policy for their development has to be participative. Tribals have to unfold their capabilities and participate in development programmes which is possible only through the institutions of Gram Sabha. The 73rd Constitutional Amendment Act, 1993 had given Constitutional status to the Gram Sabha, and State Legislature was to strengthen the institution of Gram Sabha by specifying its powers. The constitution has also provided for the establishment of an important agency of Tribes Advisory Council. These Councils advise the government regarding formulation and implementation of policies and programmes for tribal development. Tribes Advisory Council is an important forum to provide machinery with adequate consultancy status representing tribal interests. The State Governor is specially empowered to take advice from Tribal Advisory Council consisting of 17 tribal MLAs and three other official members. As per the Constitutional provisions, the National Commission for Scheduled Castes and Scheduled Tribes has also been appointed to safeguard the social, economic and political interests of tribals to prevent exploitation as per Protection of Civil Rights Act and Prevention of Atrocities Against SC/ST Act, 1989, and ensure other facilities available to SC/STs under various government programmes.

REFERENCES

Bakshi, P.M. (2010). *The Constitution of India*, Universal Law Publishing Co. Pvt. Ltd., Delhi.

Dosi, S.L. (1997). *Emerging Tribal Image*, Rawat Publications, Jaipur.

Joshi, K.C. (2011). *The Constitutional Law of India*, Central Law Publication, first edition.

Joshi, Vidyut (1998). *Tribal Situation in India*, Rawat Publications, Jaipur.

Mathew, George (2000). *Status Report of Panchayat Raj in India,* Institute of Social Sciences, New Delhi.

Mohapatro, P.C. (1987). *Economic Development of Tribal India*, Ashish Publishing House, New Delhi.

Narwani, G.S. (2004). *Tribal Law in India*, Rawat Publications, Jaipur and New Delhi.

Nehru, J.L. (1964). "Our Tribal Heritage", *Vanya Jati*, July.

Sharma, B.D., *Administration for Tribal Development*, Sage Publications, New Delhi.

Singh, K.S. (2000). *People of India*, Volume III, Scheduled Tribes, Popular Prakashan, Mumbai.

3

Ecology, Sustainability and Embedded Belief Systems

Sanjib Kakoty

Introduction

During the last couple of years, different countries of the world have borne the brunt of severe climatic conditions, be it in the form of unprecedented droughts or devastating floods, tornadoes, landslides along with major earthquakes and tsunamis. Are these natural calamities natural and normal? Or should one accept the argument that these natural calamities have been invited by our own actions and inactions that has resulted in severe climate change. Looking at the nature and frequency of numerous geo-climatic upheavals that the earth is experiencing it appears that the worst case scenario in the climate change advocates seems to be coming true.

But the question which arises is that if mankind is aware that its actions are directly contributing to climate change, why is it that corrective measures are not being taken? The answer probably lies in the socio-political and economic system created by man that requires the achievement of a perpetual growth cycle which is based on ever growing production and consumption. Unfortunately, such a growth paradigm is impossible to achieve in a scenario where the earth resources are necessarily limited. Thus, it is apparent that the problem stems from the fact that mankind does not know how to reign in its need for continuous economic growth without adversely affecting ecological sustainability.

This problem of achieving and maintaining sustained economic growth seems to be a problem that is unique to the modern post-industrial economy. It is being increasingly realized that the ancient civilizations may have economic systems and practices that was based on living in harmony with nature. Interestingly, sustainability research clearly reveals that the production and distribution systems of ancient civilizations and the philosophical basis of it, were based on what constitutes the modern day best practices of sustainability! Hence, there is a growing clamour to learn from the wisdom of the ancients and incorporate their ideas and practices or in other words learn from the philosophies that have stood the test of time. As North-East India is home to numerous traditional societies that have followed old age practices up to the modern times, this region could make significant contribution to the sustainability discourse. Tribal communities and indigenous religions the world over are accepted repositories of ancient knowledge and wisdom. Significantly, with the tribal communities, such knowledge systems are living traditions. Thus, such living knowledge systems could be an important source of guidance for modern man in his quest for sustainability.

Under the accepted parameters that scale development, skewed as they are towards measurement of GDP/GNP, modern industrial societies seem to be much more developed than traditional societies. But when one widens the scope of the discourse and include concepts such as quality of life and the happiness index, the equation could significantly alter. It is an accepted fact that after Bhutan successfully argued for the inclusion of what has been termed as the happiness quotient into the development paradigm, it resulted in a significant and qualitative change in the understanding of development measurement indices. In a couple of years, the happiness index has become an integral part of the development measurement indices being adopted all over the world and is gradually emerging as a cornerstone of the development debate.

After the concept of happiness established itself as an important component of measuring development, it was hardly surprising that other qualitative concepts soon followed. Of these, the two major concepts that slowly gained ground in developmental economics are sustainability and quality of life. While sustainability slowly emerged as the key focus of the modern world especially after the World Commission on Environment and Development known as the Brundtland Commission, published its report in 1987, and the United Nations adopted it. The quality of life debate is yet to develop universally accepted measurement indices. None the less, what is pertinent is that the concept of sustainability would necessarily encompass the quality of life component within its discourse.

When one tries to understand and explain the modern day concepts of sustainability there is a growing realization that pre-industrial and ancient civilizations could boast of practices that were eco-friendly and sustainable. Deeper examination of the production and distribution systems of ancient civilizations and the philosophical tenets that supported it, were based on what constitutes the modern day best practices of sustainability! Hence, of late, there is a renewed interest to research and learn from the wisdom of the ancients and incorporate philosophies and practices that have stood the test of time. It is in this context regions that have considerable populations of traditional communities, such as the North-Eastern region of India, could make significant contribution to our understanding of sustainability.

The North-Eastern region of India is home to a number of communities and tribes that have lived a life of isolation. This isolated existence ensured that these communities had simple economies that were essentially in a pre-industrial stage. Many of these communities were also in a pre-literate stage and had world views and belief systems that may be compared to nature worship. All over the world such communities, their belief systems and religions are recognized as being repositories of ancient knowledge and wisdom. Significantly, with most tribal communities, such knowledge systems are not a part of history but are a living tradition that continue to be practiced. Thus, such living knowledge systems are considered to be an important source of guidance for modern man in his quest for sustainability.

What traditional practices and belief systems of the people of the North-East has stood the test of time? How can these be used by modern man to achieve the goal of sustainability?

Khasi Sacred Forests

Forests have occupied an important place in the life and living of tribal communities. Not only are forests used for the sustenance of the community but it is also considered to be the abode of gods and spirits. Thus, forests were used for survival of man, and in turn man looked after the upkeep of the forests. It was a symbiotic relationship and one of mutual benefit based on respect. Not surprisingly, specific forests, in different parts of the world has been considered to be sacred by communities living in its vicinity. In the case of the North-East too, there were a number of forests that were designated as sacred. However, with the onset of modernity, most of these forests have disappeared. Some of the last of the sacred forests can be found in the Khasi Hills of Meghalaya. These constitute one of the best examples of ancient wisdom in ensuring ecological sustainability.

Not unlike the belief system of tribes in different parts of the world, forests, rivers and trees were thought to personify gods and spirits. These were venerated and worshipped, and their places of dwelling, be it forests or caves were considered as sacred. These abode of the gods were not only revered but also worshipped and protected. As a result, in spite of tremendous pressure, these forests were protected and preserved. This preservation was certainly helped by the belief system that made it taboo to pick up even a leaf and carry it out of the forest! The belief system also firmly held that any disobedience of this would have serious consequences such as ailments, which unless atoned for, through prescribed rituals, could prove fatal. In addition, it was also the normal practice to pray, pacify and seek the blessings of the forest gods, through prescribed methods. The forests spirits were also appeased through rituals and sacrifices and in turn the spirits ensured the welfare of the people.

In the modern age, such beliefs are often ridiculed. Dubbed as animism and superstition and due to the relentless pressure of more organized religions, traditional belief systems are fast disappearing. With them is disappearing a way of life that was instrumental in preserving the sustainability of human civilization. As a case in point may be cited the example of the sacred forests of the Khasi tribe. In the not too distant past, many forests of the Khasi and Jaintia hills were designated as sacred. In fact, even the name of the capital city of the state, Shillong, is derived from the hill U Lum Shyllong, where the present day Shillong Peak is located. It was considered a sacred forest, where resided the spirit of the guardian deity, U Lei Shyllong. The location of this sacred forest does not appear to be an accident as the Laitkor forest Range which housed the sacred forests, are today recognized as an important rain catchment area for the Shillong valley. It is due to this forest that most of the perennial springs that supply crystal clear water to a large part of Shillong round the year owe their existence. Today, with the denudation of these forests, numerous hill springs are drying up and water scarcity has become much more common.

Sadly today, what remains are a mere handful of sacred forests. In a landscape that once boasted of tens of such forests. Of those that survive, perhaps the most celebrated is the one located at Mawphlang. Located some 25 km away from Shillong, this forest with an area of 76.08 acres, is considered to be an ecological and biodiversity hotspot. It is assumed that the core forest area has remained undisturbed for over five hundred years and is home to various exotic species of flora and fauna. In spite of the tremendous pressure of population and ever growing hunger for resources, the only reason that the forest has survived was because it was designated as sacred forest of the Khasi faith and it is the belief system of the indigenous people that continues to protect it. This verdant forest is also considered a rain catchment area and probably, it is no co-incidence that much of Shillong's growing water supply needs are met from the nearby Mawphlang dam, fed from the nearby perennial river !

The pristine sacred forest of Mawphlang is home to an amazing range of flora and fauna that is yet to be properly documented. Home to many exotic and endemic species, one recently discovered medicinal plant is known as the Taxus Baccata. The plant has become extremely important for the medical world as it is shown to contain huge anti-cancer properties. Interestingly, in the Khasi tradition, this tree was always venerated and called the Ksesh Blei which may be literally translated as God's tree or the Sacred tree. Why was it so named? Was it because the ancients had knowledge about its medicinal properties? These questions do throw up tantalizing possibilities.

Apatani Composite Farming

The easternmost state of India is Arunachal Pradesh. Geographically, the largest state of the region, it has the lowest population density. Numerous tribal communities inhabit the state and the Apatanis are one of them. The Apatanis live primarily in the Lower Subansiri district and are acknowledged for their advanced agricultural knowledge. Perhaps, the Apatani claim to fame is derived from their famed skills in composite agriculture. Unlike their neighbouring tribes, the Apatanis practise terraced agriculture, with rice being *trhe naibn* crop. Interestingly, their terraced wet rice fields are also extensively used to rear fish. At the same time, the elevated borders or bunds, that retain water in the terraced rice fields are also used for growing finger millets, and other items. It may be mentioned that the millet plants not only bind and hold the soil but also crucially reduces weed growth. Moreover, the millet produced is greatly used for manufacturing the local brew, that is used for all religious and social occasions. It is a normal practice to ensure that fish fingerlings are released in the paddy fields and subsequently, the fish feed on the plankton and other micro-flora and insects. As they forage for food, they release micro-nutrients from the soil that is beneficial for the paddy. The fish grow along with the paddy crop and sometimes the fish is harvested twice within the year.

In addition, through a number of ingenious methods, the Apatanis have perfected the art of nutrient management of the rice terraces. The fields are fertilized through a method of composting that is derived through decomposition of paddy straw by inundating fields. In addition to the composting, a number of essential mineral supplements are added to the fields through ash. This is derived from burning of straw and other vegetation that remain in the fields and outlying areas. At times, additional manure is derived from animal droppings, rice husks and kitchen waste are also added to the fields. An ingenious way of managing solid waste and converting it into useful manure was also commonly practiced. It was a common practice for Apatani villages which are normally located at high elevations, to direct the organic sewage generated in the villages, to the outlying fields. This was a good source of organic manure, useful for both the plants as well as the fish.

The Apatani relationship with the forest also needs to be looked at. In a classic display of traditional wisdom, the community clearly realized that the success of their agriculture was dependent on the adequate and timely availability of water. They also realized that apart from appeasing the rain gods they also had to appease the forest spirits. They clearly realized that the water supply to the terraced rice fields were largely dependent on the mountain springs and streams. In order to ensure that these did not run dry, they were aware that it required the preservation of the catchment areas. Thus, the Apatani developed embedded belief systems and social customs that prohibited the cutting of certain trees such as the *Kiira* or the *Castapnopsis*. Interestingly, this tree has a unique deep root system that helps water percolation and retention. In addition severe restrictions and taboo against tree felling, coupled with provisions for exemplary punishment ensured that the trees and forests of the catchment areas were well preserved. In effect, Apatani customs and traditions display ecological sagacity in land and water management and optimizing the production process, that is certainly worthy of emulation even today.

It may be mentioned that the Apatani traditions are not unique and similar practices are seen among many other tribes. For instance, the Zabo practice of the Chakesangs of Nagaland, also bears comparison to the agricultural practices of the Apatanis. Here too, there is an integrated approach to agriculture, fishery and forestry, based on traditional wisdom that has stood the test of time.

There is hardly a doubt that the embedded belief systems of the communities of the North-East can provide important lessons for a world grappling with the problem of sustainability. Do we have the wisdom to learn from our ancestors and create a sustainable future for our children?

SECTION II

Women and Natural Resource Management Strategies

4

Naga Women as Natural Resource Managers: Is Revival of Tradition Always the Best Option?

Lucy Zehol

Introduction

Gender

In everyday conversation, however, the term 'gender' may more often be used interchangeably with 'sex' as a categorical reference, i.e. 'male' and 'female', which are two examples of 'gender'. The distinction between 'sex and gender' as analytical categories, can be traced back to developments in sociological, medical sciences, and (social or cultural) anthropology of the twentieth century. Whereas Simone de Beauvoir's (1949) political criticism of the subordination of women, titled "The Second Sex", linked the cultural devaluation of women in Western society to biological differences between men and women. Gayle Rubin (1975) explored the notion of a sex or gender system, whereby the 'raw resources' of male and female bodies become 'gendered' products through processes of socialization, normative behaviours, hegemonic representation etc.

The 'raw resources' referred to by Rubin were defined by a conception of biomedical differences between 'male' and 'female'. Research on chromosomes, gonads, differences between the genitals, hormones, brain size and anatomical structure (men have more grey matter, women more white matter) and secondary characteristics such as body fat distribution (including breasts) and body hair. Bio-medical research has consistently sought to link such differences to innate 'personalities' or psychological explanations for perceived differences in behaviour and effective experiences, whereas sociologists and social-cultural anthropologists as well as many psychologists have tended rather to focus on the ways in which symbolic meaning and practices are associated with biological features of 'sex' to 'cultural construct of gender'.

Since the 1990s, theorists have drawn attention to the fact that the medical assignment of 'sex' at birth is also a culturally constructed practice. Ruth Hubbard (1990) cites Anne Fausto-Sterling's estimation that 1-2 per cent of babies are actually born with ambiguous sexual characteristics (chromosomal, gonad, genital sex do not always corroborate with one another) and proposes the existence of at least five sexes. Like Judith Butler, Hubbard sees the surgical procedures to 'correct' the genitals of intersexed infants as an example of science being used to forced 'natural' bodies to conform to cultural perceptions.

In response to the increasing abstraction of gender theory, with everything being reduced to cultural constructions, Toril Moi's essay "What is a Woman?" (2001) argues the need to abandon the 'gender' concept altogether and return to a theory embedded within a material world. Her 'lived body' concept proposes that a person's identity interaction with experience of their world is directly related to the materiality of his/her body, its physical attributes and capacities, the reactions it evokes from others, and its 'situation' within a historically specific cultural environment. Moreover, the body's inhabitant has projects and goals that determine the way a person interacts with this 'situation'. Thus, the limits of the 'gender concept' which tend to imply that people are subject to the roles imposed upon them by 'culture' and transcended by the 'lived body' theory. The advantage of the 'lived body' conceptualization is that it accommodates a dynamic aspect to gender on the individual level that can be generalized to the members of a community as a whole within an overall historical framework. However, as Young (2005) points out in the essay "Lived Body *versus* Gender" such a conceptualization is more useful to describe individual experience than macro-logical social structure and organization. Young concludes that 'gender' continues to be necessary in political and sociological approaches to gender relations.

Gender Relation in the Human-Environment Interface Among the Naga Society

This chapter attempts to make some observations at the human-environment interface within a local context and the internal gender relation as the drivers of ecological change by applying a political ecological framework in understanding the dynamics of resource management in the context of Nagaland. Peluso and Watts provide tools for thinking about conflicts and struggles engendered by the forms of access to, and control over resources. Moreover, its attentiveness to power relations inherent in defining, controlling, and managing nature suggests an alternative way of viewing the link between environment and political action (Peluso & Watts, 2001: 24-25).

Each Naga tribe is distinct with regard to its languages, foods, dress, customs, traditions and many other things. This distinctiveness further extends to the village level with each village differing in one way or the other from the nearby villages. Given such diversity, exceptions to the generalizations made in this chapter are inevitable.

We will begin by mapping briefly the relation of Naga women and men to land and forest as these are the life support system around which the Naga economy and the knowledge system evolve.

Land

Men can inherit land which is one of the most important productive natural resources for people of Nagaland; however, it does not apply to their women. Naga societies follow the patrilineal system in which descent is traced through the paternal line. Men can inherit land and property while women are granted only user rights.

The (eldest) son inherits the ancestral property whereas a daughter has no right to inherit the immovable property from her father or claim the ancestral property. Out of love and affection, she can be given a share of the father's property as a gift. However, such declaration of gift should be made in the presence of kinsmen and in-laws as witness. A woman is entitled to use the gifted plot of land as long as she lives and she has to return it to her brothers after her death. In some villages, women can transfer the gifted property to their daughter for one or two generations after which they have to return it to the brothers. In the absence of brothers, the land goes back to the clan.

A widow can look after her family property as long as she remains under her deceased husband's roof. According to the customary laws, women in Naga societies cannot usually claim the children after divorce. Maintenance of an infant is given to the mother in case of divorce for a short period of time, which varies from village to village. Moreover, women are not allowed to participate in the Village Development Board and share their ideas regarding village administration. Women are excluded from participating in all the important decision-making institutions (Zehol, 2009).

Therefore, Naga women's access to land resources and user's rights are mediated by the Naga customary laws and control rights are often subjected to Naga customary laws. Regarding the principles relating to the relationship between the men and women shows that rules were more favourable for men as husbands than women as wives, especially where inheritance of properties was concerned. In the customary law of the village as in the other Naga villages like Khezhakeno, women had limited rights when compared to men in the political or administrative sphere though their role was seen to be very essential for the proper functioning of the society. The recognition of Naga customary laws in Article 371(A-II) of the Constitution of India further marginalizes women.

Women's lack of ownership in land especially immovable property affect economic well-being, social status and political voices and risk of domestic violence. And above all, due to their low representation in decision-making spaces, they have little say in the rules governing natural resource management now being frame shows participatory exclusion in environmental governance.

Forest

Forest is a store house of resources of various kinds like fuel wood which is an important part of a Naga kitchen, medicinal herbs as well as raw-materials needed for subsistence farming (green manure, fodder for cattle etc.). Forest also contribute to rural livelihoods through the role they play in replenishing groundwater supplies and preventing soil erosion.

The forest has traditionally been and still is a source of a wide variety of wild plants, insects and animals for the Naga people. Apart from wild animals and plants, a great variety of other Non-Timber Forest Products are commonly collected from the forest. Among these are medicinal plants, fruits and roots, bamboo and cane materials that are used in basketry and handicraft industry, wax and wood oils such as pinewood, traditionally used for lightening purposes, broom grass and bamboo shoots, widely used in Naga cuisine. Government schemes under the approval of the Ministry of Environment and Forests in some villages regulate the extraction of timber and non-timber forest products. The rich mineral resources of the State are not yet been exploited, among these are petroleum and natural gas, marble, coal and limestone (NSHDR, 2004).

Women and children are the main collectors of wild edible plants. Men on the other hand are traditionally involved in hunting in spring, winter and autumn season. Although, hunting is still practiced to some extent, many animals like fox, squirrels, deer, bear, wild boar and monkey have drastically decreased and some species disappeared, e.g. the Hornbill is among the most endangered species in the North-East region, owing to the loss of its habitat and rampant slaughtering for its meat, beak and feathers. The Hornbill is a majestic bird that is held with great esteem, especially the Great Hornbill, the largest true Hornbill, whose tail feathers are used by most Naga tribes for traditional purposes particularly on headgears. The Great Hornbill has earned itself a place in the Naga culture. In general, Hornbills also play a very important role in maintaining a healthy forest by dispersing plant seeds as they move about.

Sadly, with rapid urbanization that has resulted in mindless destruction of the natural environment, rare animal and plant species have disappeared from the face of Nagaland during the past few decades. Due to intensive timber extraction activities in the 1980s, vast areas of virgin forests were denuded. Felling of trees is now regulated wherein timber and other forest products can be extracted only through working schemes and as per the Tree Farm Rules, 2002. The conservation initiatives of the State Government, village council and local people, combined with the ban on hunting, on timber extraction and on forest burning led to increase in the overall vegetation cover and regeneration of species at all levels (NSHDR, 2004).

There is a gradual erosion of knowledge of local plants and species, as the forests and commons, the basis of this knowledge disappear. While rural Naga women are by no means the sole repositories of this knowledge, they are often the significant bearers of information on particular items such as food-related forest produce. Also while women have no direct role or much say in the community forest management, effective forest resource management and its wise utilization will provide both men and women with ecological benefits, life support systems and resources.

Gender Division of Labour

The gender division of labour is the basis for women's having a double work burden. Women's double burden of productive and reproductive work has been widely documented. Changing this is of course the most difficult but women who are economically independent and have a political voice are also in a better position to negotiate some change on this count. Basically, in order to transform the relationship between women and men and people and nature, we need to enhance the bargaining power of women in relation to men (both within and outside) the household. Women's bargaining power would depend on three types of interlinked factors – their material situation like independent economic status, economic, social, and political support from external agencies such as kin, the state and NGOs and gender ideology such as gendered norms and perceptions about women's appropriate roles, rights, abilities etc.

The gender division of labour depicts power relation and inequalities in sharing of works. Men partake in activities that need physical strength and stamina but are seasonal. Men also do activities that have economic value such as collecting timber and marketing cash crops. Women are attached to activities mainly for sustaining livelihood that are invisible and unaccounted. There is a gender gap in sharing the workload attached to resources and value. Thus, the gender differentiated roles and responsibilities in natural resource management and interventions must address the specific needs and opportunities of rural women and men, to reduce inequalities, stimulate growth and reverse environmental loss and degradation.

Women may not necessarily be the source of solutions to environmental problems. Groups of women as well as men may act in environmentally damaging ways and women's apparent closeness to nature may stem from their lack of opportunities outside the household and the social relations in society than their biological make-up. Sustainable development, therefore, depends on removing women's subordination and oppression as well as their poverty, not simply grafting on women as a group or lumping women's varied interests together.

Gender approach to the environment policy therefore, should attempt to understand the differences that exist between women and men in households and between subgroups within communities, differences in workloads and use of resources, knowledge and skills, social status etc. Besides, an attempt should be made to understand power structures and relations, thus recognizing the rights and responsibilities of women and men that exist in communities. These may be partly hidden and may not be obvious without careful investigation. For example, traditional rights of access to land and the products it yields may often be problematic for women as women may not have control over farm products and earnings. These rights might not be written or codified. The socially constructed gender division of labour with women in reproductive, productive and community organizing works give women less time to invest in their human capital or for capacity building for social networking. In general, environmental and resource degradation increases women's workload and drudgery.

The inequalities as stated above relates to three aspects in particular: (1) property rights, especially land and forest rights; (2) The gender division of labour; and (3) political authority, power and access to public decision-making spaces.

Emerging Issues

Today, there are considerable efforts by environmental academics, and activists, emphasizing the importance of local participation for successful environmental regeneration. As a result, today we are seeing an emerging consensus among scholars, governments, and NGOs that local resources should be managed by village communities. But what shape should community institutions for environmental management take? Or to what extent can it so serve? It is important to probe this, since eco-feminism is becoming increasingly important in shaping agendas on women and the environment in international and national forums and among donor agencies.

A number of influential environment thinkers – Carolyn Merchant and Vandana Shiva in South Asia favour the reviving or replicating of traditional community institutions on the assumption that the past constituted a period of ecological stability and social harmony. Termed by some as the 'new traditionalists', they represent the pre-colonial period as one of ecological and social culture. Some even consider the entire period from 500 B.C. to 1860 A.D. as one of relative environmental stability, where caste and associated occupational divisions provided a non-conflictual basis for sharing resources. In that period, communal institutions for resource management are also assumed to have been a common feature. All this, it is argued, was destroyed mostly by colonialism.

Existing evidence provides a very different picture. Undoubtedly, colonialism contributed to environmental decline, especially through large-scale logging for railways, ship-building, and plantations, and through State takeovers of forests and communal land, which undermined community property rights and village institutions. But the pre-colonial period was far from one of ecological stability, given the growth in human populations and spread of agriculture. And although communal resource management institutions clearly did exist, there is little to suggest that these were widespread nor can their erosion be attributed only to colonial policies.

Historical representation apart, what is especially problematic in this romanticization of the past is its uncritical perspective use. Moreover, the revivalist agenda is not confined to a few. According to Gadgil and Guha, in almost all parts of India, social activists of a range of ideological persuasions – "Gandhians, Marxists, and wholly apolitical social workers" – are seeking answers and wanting to revive traditions.

Notwithstanding and not denying the lessons traditions could offer on some counts, what is at issue here is the inegalitarian social relations and values that most traditional institutions embody. In the endorsement of such institutions, there is in fact a striking absence of discussion on their effect on social relations, especially gender relations. Equally, many of the community-based institutional arrangements that have emerged in recent years through the encouragement of governmental for example, the Government of Nagaland initiated the concept of communitization of Public Institutions and Services Act (2002) in order to build partnership between government and the people through delegation of management responsibilities to the community so that the performance of the public utilities improves in areas such as health, water supply, forest, education, bio-diversity, tourism etc. are found to be entrenching, even exacerbating, gender inequalities.

The gender division of labour continues to be unequal with women bearing a double work burden. Although acquired property may be allowed to be inherited by women, significant inequalities remain, especially in relation to ancestral landed property. And despite an improvement in women's education, there is still a wide gap in relation to men at all levels, but especially at higher levels of decision-making. Nagaland has not seen even one woman candidate elected for assembly election till date. And particularly due to their low representation in decision-making forums (lack of critical mass of women), they have little say in the rules governing natural resource management now being framed. This is a very central argument, that although is understood by self-respecting development workers and bureaucrats dealing with development, unfortunately not often brought into actual policy (Kirpal, 1995).

This neglect of gender in most environmental writing and policy continues, despite the growth of eco-feminist literature. In fact, rather than challenging the traditional view, the premises of eco-feminist tend (however unwitting) to reinforce it, and thus provide support to institutions and practices that can sustain (or even increase) gender inequalities. Even in the global arena, largely due to eco-feminist advocacy, while the importance of women's role in ecological regeneration is beginning to be recognition, the ground for this recognition is commonly the view that women have a special way of knowing and nurturing nature. So, what policy-makers pick up is the idea of women's naturalized roles, not rights.

Observation

Rendering hierarchical structures historically invisibly make invisible their continuation in the post-colonial period. Today, these forms of inequalities persist, although in less stark forms. This is a need for all of us to consciously transform the formal structures that control natural use and abuse including existing socio-cultural and political barriers which seems to act as a double edged sword for womenfolk in particular. It is in everyone's interest that the earth is green but it is equally important that women are not just expected to contribute to the green but must be able to claim equal right to it in dignity and honour.

REFERENCES

Beauvoir, S. (1949). *The Second Sex*, Translated by H.M. Parshley 1972, Canada: Penguin Books Ltd.

Butler, J. (1990). *Gender Trouble: Feminism and the Subversion of Identity*, New York: Routledge Publications.

Escobar, A. (2008). *Territories of Difference: Place, Movement, Life, Redes*, Durham, NC: Duke University Press.

Gadgil and Guha. (1995). *Ecology and Equity*, New Delhi: Penguin Books.

Hubbard, R. (1990). *The Politics of Women's Biology*, New Brunswick: Rutgers University Press.

Kirpal, P. (1995). *Culture and Development*, New Delhi: Har-Anand Publications.

Marion, Y.I. (2002). "Lived Body *vs.* Gender: Reflections on Social Structure and Subjectivity", *Ratio*, 15: 410-428.

Marion, Y.I. (2005). *On Female Body Experience: 'Throwing Like a Girl' and Other Essays*, Oxford University Press.

Mercant, C. (1980). *The Death of Nature: Women, Ecology and the Scientific Revolution*, San Francisco: Harper & Row.

Mies and Shiva, V. (1993). *Ecofeminism*, Delhi: Kali for Women.

Moi, T. (2001). *What is a Woman?*, New York: Oxford University Press.
Nagaland State Human Development Report (NSHDR) (2004). Department of Planning & Coordination, Government of Nagaland.
Peluso, N.L. and Michael, W. (2001). *Violent Environments*, Ithaca & London: Cornell University Press.
Rubin, G. (1975). "The Traffic in Women: Notes on the Political Economy of Sex", in R.R. Reiter (ed.), *Towards an Anthropology of Women*, pp. 157-210, New York: Monthly Review Press.
Shiva, V. (1988). *Staying Alive: Women, Ecology and Survival*, London: Zed Books.
Sinha, S., Gururani, S., and Greenberg, B. (1997). "The New Traditionalist' Discourse on Indian Environmentalism", *Journal of Peasant Studies*, 24(3): 65-99.
Zehol, L., and Kevekha K.Z. (2009). *The Legendary Naga Village – A Reader* (Khezhakeno), Dimapur: Heritage Publication House.

5

Women and Forest Resource Management: Livelihood Strategies and Interventions

Rekha M. Shangpliang

Introduction

Natural resources such as land, forest, agricultural crops, water, timber, wildlife, and mineral resources form the most basic needs of a large majority of the world's population and are the driving force behind economic development, industry and Gross Domestic Product (GDP) in any country. Of late, there has been a growing trend of focusing thrust areas of development towards decentralization of community resource management. This trend was visible in the early 1980's in the whole of South Asia in particular when Central and State Governmental agencies including development practitioners were moved by the changes taking place in the environment-man relationship. Depletion of the resource base, livelihood strategies of rural folk, centralized system of resource management and the counter-attack of communities through social movements were some of the issues that called for a deeper understanding of the resource management systems.

Women's centrality to resource management can hardly be underestimated. The traditionally demarcated roles that women perform in their domestic domain such as washing clothes, collecting firewood and farming, tending animals, growing vegetables and collecting Non-Timber Forest Products (NTFP) are considered 'natural' and as such contribute greatly to resource conservation. The conserving and sustaining roles that women perform has been glorified in the pages of history to the extent that women are valorised as 'resource managers'. No one can forget the historic "Chipko Movement" that took place in 1973 in

Mandal, a small hamlet at Chamoli district of Garhwal (now in Uttarakhand) where women activists staged protests against tree felling by literally 'hugging' the trees. That was the starting point of a conservation endeavour that was 'pro-poor' and above all 'women-centred'. What message did the Chipko Movement have for its forerunners? Firstly, the word 'Chipko' itself which means 'to hug' in Hindi evoked pathetic images of poor village women desperately trying to protect their forests from the axes of greedy contractors, by hugging the trees, which was their only means of economic survival. Secondly, Chipko has become an environmental icon to the world today that inspired 'eco-feminism' with a strong message "Our lives before our trees".

To many of us, the landmarks of the Chipko movement may have left behind a legacy of a well fought battle, a women-centred movement for natural resource conservation, but today 46 years after its inception, questions remain: What role do women have in natural resource management today? What are the challenges that constraint them from their accessibility to forest resource? Let us start first by understanding the inextricable link between natural resources and rural livelihoods, in particular the increasing role of tribal women as household providers in declining rural economies. Tribal economy is inextricably linked with forest livelihood wherein a major source of their income is derived from dependency on forest-related activities. However, what is distinctive about tribes in India is their poor socio-economic condition. Poverty has been one of the most serious problems plaguing the country. Studies have also revealed that dependency on forests and common property resources increases as a tribal household becomes economically marginalised. And there is no denying the fact that as poverty increases, the tasks of women become more prominent in ensuring the survival of households by assuming greater responsibility to provide resources from forests and common lands. The importance of women in collection of forest produce is borne by data from almost every country in Asia, Africa and Latin America. For instance, a study in North-West Frontier Province of Pakistan showed that 78 per cent of morel mushrooms are collected by women and children, which have supplemented their daily requirements for food and other basic necessities. Similarly in West Bengal, tribal women gather *Sal* (*Shorea robusta*) leaves for six months in a year and earn about ₹ 72 per month under the best circumstances. Like most of the tribal Indian women many Naga women depend on the forest and its resources for their livelihood, spending most of their labour and time tending their agricultural field and foraging the forest to meet the subsistence needs of their families and also for income. They have a very close relationship with the environment they live in and can be called the "guardians of their biodiversity" and caretakers of most agricultural and livestock resources (*Nagaland Environment Protection and Economic Development and International Institute of Rural Reconstruction*, 1999, p. 168).

Khasi Women in a Matrilineal Society

The Khasi are a matrilineal tribe inhabiting the Khasi Hills in the North-Eastern State of Meghalaya. While there are many interpretations of the origin of the Khasi race, the term 'Khasi' is believed to have derived its meaning from the terms '*kha*' meaning 'born of' and '*si*' meaning 'ancient mother', which when interpreted means '*born of the mother*'; thus bringing out the matrilineal character of the Khasis who trace their descent from the mother. The Khasi follow the principles of matrilineal descent, where ancestral property is passed from the mother to the daughter, preferably the youngest daughter or the '*khadduh*'. The matrilineal Khasis of Meghalaya believe in the ideology – "*Long jait na ka kýnthei*", which means "From the woman sprang the clan". This ideology is so deeply rooted in the Khasi ethos that it has brought to light the role of women in perpetuating the clan from one generation to another. Descent line in a Khasi family is reckoned only from the mother's clan or '*kur*' as a result of which the children belong to the descent group of the mother. Therefore, it is customary for them to speak of a family of brothers and sisters who are great grandchildren of one great grandmother, and identify themselves as '*shi-kpoh*', which literally means 'one womb' "that is the issue of one womb". Significantly, the Khasi ideology of human reproduction describes the father as the provider of stature and form (*U kpaubaaika long rynnieng*), while the mother contributes flesh and blood (*kakmiekabaaiiakadohkasnam*) to the child. This filial bond between the mother and the child is strongly reflected even in familial relations and kinship ties where the term *parakur* (mother's kins) is used more significantly than the word *Bakha* (father's kins).

A woman is central to the family in Khasi ideology. She has the responsibility of nurturing and continuing the matrilineal descent line through reproduction. The Khasi woman has numerous familial and kinship roles to perform. Besides being the custodian of family property the 'Khadduh' or youngest daughter's role is all the more heavy and burdensome. She has to look after the aged members of the family, and other members of the matrikin, she inherits the family property and continues tradition of carrying out the religious rites and ceremonies from time to time although with the help of her mother's male relatives. The Khasi belief in the saying "*Ka kynthei ka bat iaka niam*" meaning, "It is the woman who holds the religion", which signifies her responsibilities in the religious domain. However, the exclusion of Khasi women from having any say in village durbars and meetings is noteworthy. They are not allowed to speak aloud in public meetings and village council meetings perhaps because of her domesticated role and nurturing skills she is expected to confine her duties to the home and family matters, leaving the others to the male.

The Khasis have a deep-rooted affinity with nature. The earth which is symbolized as 'mother' or 'meiramew' is looked upon as the caretaker of all the natural resources. Thus, forest holds an important place in the social, economic and religious life of the Khasis. It is looked upon, as a well loved home, a game sanctuary and an abode of worship all rolled into one, around which their social, cultural and religious activities revolve. It is a familiar sight to see women and children setting off into the woods to collect edible fruits and roots. A typical day out for an average Khasi village woman would be best described as follows:

> "She carries her baby on her back, climbing the hills she sets off for the woods with a *ja-song* (cooked rice packed in a leaf) salt and some dry fish (*ktung*). Going to the forest virtually takes her the whole day so besides collecting twigs and fuel wood and other forest products, she also spends the day washing clothes on the riverside in the foothills" (Shangpliang, 2012: 26).

As nurtures of family line, Khasi women have had a significant role to play in the domestic sphere (Uberoi, 1993: 180). Her familial roles are well defined and her 'glorifying status' as a mother has furthered her roles and responsibilities towards meeting the immediate needs of the family. These roles include some burdensome duties and responsibilities, which are part of their household chores such as carrying water, fetching firewood, washing clothes, collecting twigs and edible plants and roots which are carried out in the jungle skirting the village. This has partly been borne out of the fact that women have had a closer interaction with nature and this has enhanced the emotional bond with nature since it has helped them cope up with a number of economic hardships.

Khasi Rural Women and Indigenous Ecological Wisdom

From a gendered perspective it is a well-known fact that women's indigenous knowledge of eco-systems has contributed to ecological sustainability, and is illustrative of the strength of their eco-system-based learning and commitment to non-violative uses of nature (Wickramasinghe, p. 183). A cursory glance over the Khasi rural eco-system reveals the extent of relationship between women and ecological sustainability. Their wisdom and knowledge about certain species of edible and non-edible plants, food culture and regenerative process of soil fertility has been extremely instrumental towards judicious use of the resources. A study conducted by the author in Lawbyrwa village of Ri-Bhoi District of Khasi Hills revealed that 80 per cent of the time spent by the women of this village in a day was mainly to fulfil the subsistence needs of their family and these activities included collecting fuel wood, water, edible roots and tubers, broomshrubs and other Non-Timber Forest Products (NTFP). Since most villages are located in the forested regions and fringe forest areas they have properly defined gendered work patterns and responsibilities. While men are actively involved in agriculture, daily wage labour and farming activities, women are seen to spend more time in the woods and are the main providers of basic needs such as fuel wood, food, medicinal and Non-Timber Forest Products to the family.

There are many Khasi folktales, and legends that have played a very important role as narratives in focusing the close relationship between women and nature. The Khasi folklore and legends, "*Khanatang bad Puriskam*" are mostly woven around the various forces of nature such as the hills and vales, rocks and caves, the flora and fauna. These elements of nature are personified in the legends as the mother and son, husband and wife and friend and foe according to their natural behaviours, their love, hate, jealousy, pride and vanity which are projected out before the listeners with the sole purpose of teaching moral and spiritual values. The ecological wisdom of the Khasis is a wisdom that is solidly based on 'experience' and they depend on this knowledge to fulfil most of their day to day needs (their economy). Through agriculture, hunting, fishing and gathering a huge range of forest foods and materials with which they make most of their articles of daily use, from houses to ploughs, baskets to leaf-plates. Majority of the Khasi households are built with thatch roof, plank floor, bamboo or stone walls which are all obtained from the forest. Fire always keeps burning on a stone hearth in the centre while the smoke finding its exit as best as it can. The amount of firewood burnt by each family is anonymous. They possess an ancient lore of herbal medicine as well as knowledge about many aspects of nature and human life. Much of this ancient wisdom is today the result of oral tradition. Forest wisdom has enabled the Khasis to live in harmony with nature besides elevating them to such a high pedestal that no animal or living creature could compete with.

It is also interesting to note the use of simple objects of nature in their day to day activities that reflect the importance of nature in their cultural and religious life. The naming ceremony of a new born infant '*Kajer kathoh*' is an important occasion for the mother and the whole family where the pounded rice flour is placed on a bamboo winnower called '*Kaprah*'. When an expectant mother delivers her baby, a sharp splinter of bamboo is used to cut the umbilical cord. No metal objects or even a knife can be used on this occasion. When the umbilical cord, after being tied falls off, a ritual is performed by offering worship to certain water deities or "*Ka Blei Sam Um*". Singing and dancing form an important component of Khasi culture. There are many Khasi traditional dances that are of ecological significance – harvest dances that are performed during harvesting, another well-known dance is "*Ka Shad Kynthei*" where only women take part in the religious dance invoking God and thanking him for a good harvest. There are also dances exclusively meant for women connected to fertility. It is called "*Ka shad Nohsakyriat*", which is performed by the women folk who spin themselves on a traditional pole round and round in order to symbolize their strength and feminity of motherhood.

Forest Livelihood and Khasi Rural Women

Like most tribal communities, land and forest are the two most important natural resources to the Khasis. However, they have to be dealt as separate entities. Land locally known as '*Ri*' by the Khasis has a deep attachment to their pattern of social organization and permeates every aspect of their socio-economic life. Land to the Khasis is a "gift of nature" that belongs to the community, therefore, access to land not only ensures economic security for the individual, but control over it symbolizes territorial integrity for the community as a whole (Nongbri, 2003).

One of the most important sources of livelihood of the Khasis is the forest which is their well loved home, an abode of worship and a storehouse of their daily necessities of life, like food, water, fodder and fire-wood. The Khasis assign a deep sense of awe and reverence to the different elements of forest in their natural abode, which conjure up to them as matters of life and death. This symbiotic relation between the Khasis and the nature has been appropriately defined by a Khasi author H.O. Mawrie in the following words – "*U Khasi u im bad ka mariang, bad ka mariang ka im bad u*", (Mawrie, 1978: 78) which means – A Khasi lives with nature and nature lives with him. The Khasi rural women venture out into the forests and engage themselves everyday cutting fuel-wood, and collecting fruits, vegetables, broom grass etc. drying and processing them and, at the fall of the day, return back home with heavy bundles on their heads. The hill slopes and deep gorges remain, throughout the year round, covered with wild banana trees, cane and bamboos, innumerable varieties of edible and non-edible mushrooms, varieties of tubers and succors etc. The tender bamboo shoots are one of the tribal delicacies consumed as a curry or pickle after processing with mustard oil and spices. Every Khasi stall located along the Shillong-Guwahati National Highway, sell these bamboo shoots and other wild fruits and vegetables gathered from the forests. The Khasi Hills forests are home to many unknown plants and herbs of great medicinal value with which Khasis have developed an indigenous system of medicine that is widely used among them as an alternative to the costly allopathic medicine. The wild grown broom grass collected by the Khasi women and children are bound into bundles of broom-sticks and sold locally or exported out of the state which fetch ₹ 20 to 50 per each broomstick depending on the market and the quality of workmanship. Some of the NTFPs gathered from the forest of Khasi Hills such as cinnamon, sandal-wood, pepper, fruits and plants and herbs are of very high economic and medicinal values in the nearby urban markets.

Women Involvement in Forest Management

Of late, the Government of India has introduced a number of schemes through which regeneration and development of degraded forest are sought to be rejuvenated for the augmentation of the forest resources and development of environment. It is imperative that the institutions responsible for forest management for creating livelihood opportunities must work closely with women, since the womenfolk in Meghalaya are major players in this aspect. While it is understood that the strength of the village community among the Khasi is their 'social capital', where all members of the family are actively involved in building up subsistence economy, the women play a singular role as gatherers of NTFPs. Thus, the policy framework must give an emphatic thrust on the involvement of women in planning, implementation, monitoring and review of any activity (Darlong *et al.*, 2008). However, the fact of the matter remains that during the past five decades of Indian planning, women have been thoroughly marginalised in the process of development, implementation and reorientation. Needless to say, the constraints faced by the women in controlling and managing land and forest resources have to a certain extent deprived them of their claims on sustainable livelihood. The scenario in respect of the Khasi women who claim to belong to matrilineal society is no better either.

In the case of community forests (*Khlaw Raid* and *Khlaw Adong*) greater autonomy in management and control of forests lies with the traditional institutions, clans and authorized individuals. In matters relating to forest upkeep and decision-making, the Khasi women have no right whatsoever. The village council (*Durbar shnong*) which is traditionally empowered to deal with the overall development works of the village does not recognize a woman as its member nor does it entertain any views or opinion from her. Problems faced by women are mostly voiced through the adult male members of their households.

Table 5.1 throws some light on the indigenous strategies evolved by Khasi women in various areas of social, economic and cultural life that have indirectly contributed towards natural resource management.

Table 5.1: Distribution of Work between Male and Female in Forest Related Activities

Sl.No.	*Activities*	*Sex*	*Number of Persons Engaged*	*Percentage to Total*	*Remarks*
1.	Cultivation and labour etc.	Male	41	22.78	45%
2.	Making Charcoal	Male	40	22.22	
3.	Collecting Firewood	Female	37	20.50	55%
4.	Collecting NTFP	Female	31	17.25	
5.	Selling Vegetables/Fruits etc.	Female	31	17.25	
	Total		**180**	**100.00**	**100%**

Source: Ph.D. Research Work, 2007.

Table 5.1 shows that approximately 55 per cent of the rural Khasi womenfolk are engaged in forest activities while 45 per cent males are engaged in non-forest activities. This reveals the preponderance of Khasi women engaged in forest as their primary source of livelihood.

Khasi Rural Women and Forest: Interventions between Forest Rights and Management

In the matrilineal Khasi society, the issue of equitable distribution of benefits in indigenous management system has often been challenged. Though in principle the matrilineal rule of female inheritance and descent is followed, but customary practices are not gender egalitarian, as authority is vested in the mother's brother (maternal uncle). Women can inherit property such as land, but they lack the power to manage it. The woman is the focal point of the household organization, but when it comes to the allocation of rights over land, a sharp distinction is maintained between 'ownership' and 'control'. While ownership of land is transmitted through women, control invariably lies with men (Nongbri, 2003: 199).

Making the plight of women worse is the system of forest management operating in the Khasi Hills. Here, the management of forest lands falls under three agencies:

(1) Autonomous District Council (Green blocks, Protected forests, unclassed forests, private forests owned by clans, domestic groups, individuals and organizations).
(2) State Government (Reserved Forests or National Parks).
(3) Community (*Khlaw Raid* or community forests and *khlawadong* or community protected forest).

The diarchical form of control and management of forest in Meghalaya by the State Forest Department and the Autonomous District Councils has contributed to dual forest policies, lack of co-ordination and division of accountability. Both the authorities formulate their own forest policies and follow a separate set of forest rules and regulations in their respective areas of operation. The main reason for such a diarchy is that, while the State Forest Department maintains a well-qualified team of skilled technical staff, it has a comparatively meagre area of 1127.23 sq.km. of forest under its jurisdiction. The District Council on the other hand, has a large forest cover of 6250.68 sq.km. under its management and control and yet it suffers from inadequate technical manpower which has resulted in the lack of enforcement of rules. Forest being the principal source of revenue, the District Councils could not afford to stop forest operation under prevailing fund constraints. The private forest owners who own substantial areas of forests indulged freely in sale of timber from their forest. A cursory glance at the available statistics of Meghalaya shows a steep rise in the process of forest exploitation resulting in sharp fall in the area under forest cover. Out of Meghalaya's total geographical area 22,429 sq.km. (23,06,069 hectares), she now has only 9,41,823 hectares of forest area which comes to barely 42.03 per cent (2001 Census).These issues have had a serious impact on people's livelihood thereby leading to infringement on their traditional rights over forest resources.

In the case of Community Forests (*Khlaw Raid* and *Khlaw Adong*) greater autonomy in management lies with the traditional institutions, clans and individuals. As per customary practices, only the permanent residents of the villages have the right to access and use of community forests and other common property resources. They can collect fuel wood, cut trees for construction of houses, collect wild vegetables, orchids and medicinal herbs and quarry sand and stones from permitted sites (Nongkynrih, 2006: 55).

It is here that one finds an invisibility of women's role in forest management and decision-making. Although women shoulder heavy economic responsibilities and spend long and tedious hours toiling in the field, trudge large distances through rugged forests in their struggle to meet the survival needs of the family, they have little to say in village administration or in the management of their natural productive resources. Many times women become silent observers of the changes around them with little or no voice in decisions taken at the council meetings. Their problems are mostly voiced through the adult male members of their household. Added to this is the problem of infringement of land ownership. There has also been an unhealthy trend of gradual shift over of *Ri-raid* (community land) to *Ri-Kynti* (private ownership) as large tracts of land and forest has been taken over by a significant decline in the quality of communal land and forest but also marginalised women's role in the use of common property resources.

While delving on the issue of the decline in the quality of communal land and forest, Supreme Court Order on Timber Ban has also had an effect on women and forest livelihood. On December 12th, 1996, the Supreme Court passed its historic judgements on the ban of felling of trees, removal of timber and wood based activities in the region. While the judgement had a conservationist aim in mind, it caused unexpected miseries to the tribal population who were totally dependent on forest livelihood. The effect of the ban fell heavily on women by imposing an additional burden on them to meet the economic needs of the family and increased their domestic chores manifold (Nongbri, 2003: 18989).

Agriculture and sale of forest produce used to be the mainstay of the rural Khasi economy. Over the years, with the decline of forest cover, the Khasi rural women have been forced into local and migrant wage employment. Livelihoods today have to be balanced between availability of forest produce, landownership and wage employment. As forests have dwindled, women still depend upon the remaining forests as supplementary source of livelihood for survival, especially in the lean summer months. They have to walk longer distances to the forest for collection and carrying back the produce as headloads. For marketing these products, they are bound to depend on bus services which are quite irregular and poor due to inaccessible roads to the interior villages. This leaves the marketing and sale of produce to the men folk. This has reduced the women's transport burden and time pressure, but the negative consequence is that they have lost control over the income from selling the firewood and NTFP that they collected. This is further compounded by the fact that the women are primarily involved only in the lower stages of the value chain such as collection, cleaning, processing and sorting of forest produce, but they lack access to or control over marketing of the produce which is mostly monopolized by men. Despite their important role in the value chain, men's marginalization has led to an increase in the burden of economic hardships thereby bearing upon the brunt of poverty.

Conclusion

The conflict of interest between the Government and the tribals over the forest wealth that was begun by the Colonial rule continued unabated long after India's independence. The Government of India tightened up its grip over the control and management of forest through introduction of a new set of laws and rules while the traditional tribal rights over land and forest of the Khasi was slowly eroded. This impacted on the Khasi rural women who were the main actors in the field.

While it is a fact that customary laws with regard to land and forest and the working plan of the government have all come in the way of women, it would not be wrong to conclude that women in particular and indigenous people in general have seriously been affected by a number of issues which revolve around the use of resources and the struggle for rights over these competing issues. The conclusive evidence thus highlights the following harsh realities faced by the Khasi rural women who are dependent on forest livelihood.

Firstly, there has been a visible transition in Khasi society with regard to the state of dependence on forest resources like food, fodder and shelter to other means of living. The depletion of land and natural forest cover which was an important base for their primary economy has resulted in paucity of forest lands for livelihood. It is interesting to note here that the worst victims are women who not only have to bear the brunt of depleting natural resources but also the subordination of their gender.

Secondly, we find an invisibility of women's role in forest management. This is largely due to prevailing Khasi customary laws where women have little to say in village administration or even management of their natural productive resources. Added to this is the unhealthy trend of a gradual shift over of *Ri-raid* (community) forest lands to *Ri-kynti* (private) landowners thereby marginalizing women's role in the use of the Common Property Resources (CPRs).

Thirdly, we are often lead by the 'self-perception' that women are merely 'helpers' and not 'workers' when it actually comes to demarcating the flow of work within and outside the household. Activities like collection of NTFP, cleaning, sorting processing of forest products are often considered to be part of their household chores and therefore, they are not 'work' at all. In many remote villages women still lack access to marketing and sale of the produce which is mostly monopolized by men thereby leading to an increase in the burden of economic hardships and losing control over the income.

Fourthly, with the current rate of 64 per cent literacy in Meghalaya and the meagre economic development that has so far been achieved, the worldview of the womenfolk in rural areas of the state has remained as traditional as in the past.

All this leads one to agree that the environment today is facing a crisis amidst stringent legislations of the Government coupled with problems of development and conflict over resource base, there has been a gradual shift in women's role as basic providers of subsistence needs and resource utilization thereby leading to invisibility of their livelihood strategies and marginalization of their rights over NTFP.

REFERENCES

Darlong, V.T. (2002). "An Overview of Forest Policies and Legislations *vis-à-vis* Forest Resource Management in North-East India", in B. Datta Ray and K. Alam (Ed.), *Forest Resources in North-East India*, New Delhi: Omson Publications.

Gadgil, Madhav and Guha, Ramchandra (1992). *This Fissured Land: An Ecological History of India*, Delhi: Oxford University Press.

Government of India Press, Indian Forest Act, 1878.

Gurdon, P.R.T. (1975). *The Khasis*, Delhi: Cosmo Publications.

Khongsit, S. (1999). *KibaNgiKhotJa*, Shillong: Mrs. SucilaKhongngain, San Mer.

Krishna, Sumi (2004). *Livelihood and Gender: Equity in Community Resource Management*, New Delhi: Sage Publications.

Kumar, N., N.C. Saxena, Y. Alagh and K. Mitra (2000). *Poverty Alleviation through Forestry Development*, Country Case Study, Operations Evaluation Department, Washington DC: The World Bank.

Mawrie, L. Barnes (2001). *The Khasis and their Natural Environment*, Shillong: Vendrame Institute Publication.

Nongbri, Tiplut (2003). *Development, Ethnicity and Gender: Select Essays on Tribes in India*, New Delhi: Rawat Publications.

——— (May 2001). "Timber Ban in North-East India: Effects on Livelihood and Gender", *Economic and Political Weekly*, 36(21), 1893-1900.

Nongkynrih, A.K. (2006). "Who is in? Who is out? Equity and Customary Community Forest Management in Meghalaya, India", in Sanga Mauhanty Jefferson Fon *et al.* (Ed.), *Hanging in the Balance: Equity in Community Based Natural Resource Management in Asia*, Thailand: RECOFT.

Shangpliang, Rekha M. (2010). *Forest in the Life of the Khasis*, New Delhi: Concept Publishing Company (P) Ltd.

Sinha, A.C. (1993). *Beyond the Trees, Tigers and Tribes: Historical Sociology of the Eastern Himalaya Forest*, New Delhi: Har-Anand Publication.

The United Khasi-Jaintia Hills Autonomous District (Management and Control of Forests) 1958, 1960 Acts and Rules, 1995. Shillong: Khasi Hills Autonomous District Council.

Tucker, Richard P. (1988). "The Depletion of India's Forest under British Imperialism: Planters, Foresters and Peasants in Assam and Kerala", in Donald Worster (Ed.), *The Ends of the Earth*, USA: Cambridge University Press.

Uberoi, Patricia (1994). *Family, Kinship and Marriage in India*, Delhi: Oxford University Press.

SECTION III

Environment and Community Life

6

Enhancing Lives through Environmental Conservation: A Short Narrative of Khonoma Village

Hormila Zhingkhai

Introduction

Environmental elements like the land, river systems, forests, mountain ranges etc. have always played a significant role in the making and shaping of society and its history. It was broadly assumed that, the environment, typically in the form of climate and topography, and sometimes in the form of disease or other 'natural' hazards, dictates the physical and mental characteristics of a society, its modes of subsistence, its cultural life and political institutions. It even determines whether a society is able to scale the heights of civilizations or is confined to the depths of savagery and barbarism (Arnold, 1996: 10). At the same time, humans transform and alter their environment according to their convenience which in consequence has an impact on their social and economic life. This fact is pertinent to the present human society, especially with regard to the environment-related issues such as climate change, global warming, deforestation, degradation of land, conflict over land, water and forest, and development and livelihood issues that we are undergoing now.

In India, we are facing diverse environmental issues such as deforestation, degradation of land, water scarcity and floods, famine, environmental pollution, bio-diversity loss, displacement of people from their ancestral land and home, loss of livelihood etc. On the other hand, Chipko Movement (1973), The Silent Valley (1978), Jungle Bachao

Andolan (1980s), Save Ganga Movement, Narmada Bachao Andolan (1985) etc., are some of the Indian stories of people fighting back to protect their forest, land and environment and regain their rights to their land, forest and water. The Environmental Conservation initiative by the Khonoma community can be said to belong to the latter category.

A Brief Introduction to Khonoma Village

Khonoma village is situated in Nagaland, one of the North-Eastern states of India. The village is inhabited by the Angami-Naga tribe and has a population of 2,500 people (Khonoma Village Development Board, 2011). Geographically, it is divided into three local residential units called *Khels*, namely *Merhumia Khel, Semoma Khel* and *Thevoma Khel*. Each *Khel* is occupied by different clans and sub-clans, and the *Khel* is the unit on which the social, economic and political life of the village is based.

Agriculture is the primary occupation in Khonoma and forest-based activities such as basketry, weaving, woodwork and carpentry, stone masonry etc., forms an integral part of their livelihood strategy. The village is administered by the Khonoma Village Council which is supported by the Khonoma Village Development Board, the different *Khel* Unions and various organizations such as the Khonoma Women Organization, Khonoma Youth Organization, Khonoma Student Union and the different Age Groups.

Like most of the indigenous communities all-around the world, the people of Khonoma also have an intimate connection with their natural environment as nature forms an integral part of their social, economic and religious life. Accordingly, their folklores, myths, ethics, ethos and way of life imply their close affinity with nature. The name of the village itself implies the significance, as the name 'Khonoma' is a derivative of a native plant called *Khwüno* (*Glouthera fragrantisima*), a plant which grows abundantly wild in the hills of the village, and a large number of which are said to have been cleared from the site selected when the village was first built. Their traditional economy, agriculture, is also based on the land and its surrounding environment. Moreover, wild vegetables, wild fruits, games of wild animal and fish forms an essential component of their dietary habits. They also collect large stacks of firewood for cooking and keeping their houses warm. Apart from these, timber for house construction and furniture, raw-materials for basketry, wood works, medicinal plants, herbs etc., are also collected from the forest for their necessary utilizations. Nature was, thus, a home, providing them with shelter and protection and all the necessary resources to sustain their lives.

Besides, for the people of Khonoma, the land, rivers and forests are not just an economic base but these environmental elements are "intimately bound up with their cosmologies and identities as communities, and as people" (Colchester quoted in Souza, 2001: 13). They consider their land as a symbol of their identity and unity that holds the family, clan and tribe together (Longchar, 1995: 71). Even their traditional religious system was embedded with the world of nature as their ancestors were worshippers of nature who believed that the land, forests, trees, rivers, mountains and hills were the abode of the spirits and all the plants, trees and creatures are considered as an integral part of their life.

Such close association with nature has led the people of Khonoma to establish a culture and way of life based on their natural environment and endowed them with a sense of eco-consciousness which has aided them to ingeniously manage the natural resources and conserve the environment. Their close association with their environment gave rise to a well structured ecological knowledge and usage that is very much linked to an engaging day-to-day experience and survival needs. They adopted various traditional practices such as never clearing the forest completely or cutting down a tree completely, using natural method of fertilization, restricting unnecessary wastage of natural resources, collecting and utilizing only the amount of products that is required, prohibiting hunting during the breeding seasons etc. Even their daily routine was based on the seasonal and agricultural cycles and their festivals and folk songs and dances are an expression of their relation to nature.

However, the abrupt contact and interaction with other societies especially with the coming of the British officials and American missionaries led to changes in their social, cultural and economic life which affected their beliefs regarding nature and environment. And this has a bearing on their traditional practices of environment and natural resource management. In particular, excessive logging and hunting as a consequence of transformation from a subsistence economy to money-oriented economy resulted in large scale deforestation and gradual loss of bio-diversity which was directly or indirectly effecting their environment as well as their livelihood.

The Story Behind the Conservation

Khonoma village is renowned for its rich bio-diversity and for their traditional practice of conservation in relation to their land, water and forest. Since the time of their ancestors, their social, economic, political and religious life was based on nature, as such they found ends and means to utilize the natural resources efficiently and conserve their environment. Such close association with nature also gave rise to a well-structured ecological knowledge and usage that is very much linked to an engaging day-to-day experience and survival needs, thereby, ensuing in a sustainable use and management of their environment and its resources. But with the change of time, and change in their social, economic, religious and political life, their beliefs and association with their environment changed. These changes, especially the change from a subsistence economy to money-oriented economy led to a circumstance where the people were very dependent on the natural environment for their livelihood, hence they have developed an intricate linkage with the environment, but on the other hand, there was indiscriminate exploitation of the natural environment because they do not have many options other than to depend mostly on agriculture and the natural resources for their livelihood. Large scale logging, hunting and excessive collection of wild products for sale led to massive deforestation, loss of the rich bio-diversity of their land, rapid depletion of the natural resources which was effecting not only the environment but their survival as well. As such, the people of Khonoma were in a dilemma where they were left with the choice of continuing with the overexploitation of the natural resources and earn quick money or to control the excessive exploitation and conserve their environment. The people of Khonoma being aware that their life and survival of the community depend on the land, water and forest decided to safeguard their natural heritage by stopping the exploitation of the natural resources and conserving the environment for themselves and the future generations.

The first step of conservation was taken with the establishment of Khonoma Nature Conservation and Tragopan Sanctuary (KNCTS), in 1998 by the ecologically conscious people of Khonoma to protect the endangered Blyth's Tragopan and other wildlife and rare plant species. Next, they enforced total ban on logging in the year 1996 and total ban on hunting in 2000, which met with heavy resistance at the beginning because many of the people were dependent on it for their livelihood.

As one of the active member of conservation mentions, "*Before the implementation of total ban on logging and hunting, on an average around 2000 CFTs of timber reached Dimapur everyday as the timber from Khonoma was of good quality and there was a high demand for it. And because there was large scale unemployment in our village, many of the unemployed youths started engaging in the logging business as this was the only means for them to earn their livelihood. Moreover, it was a very lucrative business and those who were engaged in this business were progressing and bought two-wheelers as well as cars, music systems, Television sets, started building houses etc.*" The respondent also stated that "*earlier our village came to such a position that we became famous for hunting. Even children went to the forest with air-guns in the morning and return back late in the evening with hundreds of birds and animals that they had hunted in the forest. In fact, the Supermarket in Kohima was filled with wild animal meat from our village*". It was also mentioned that in a year more than 100 air-guns (which was usually used for hunting) were bought by the people of Khonoma from the gun dealers in Kohima. Moreover, the people of Khonoma celebrate a festival named *Monmi* which is related to hunting. During *Monmi* festival, the girls goes for fishing and the boys goes for hunting, later in the evening they gather all the fishes and wild games they have collected that day and share it among themselves and give it to the elders of the village. Basing on all these practices, the villagers started opposing the ban on hunting saying "*hunting is our tradition. It is a sport for us*". Consequently, there was much opposition, arguments, conflict and even Underground militant intervention as they used to receive tax from the logging business. Some of the youth (around 300 youths) who were engaged in logging business and were against the ban even left Khonoma and went to Dzuliekima, which is an offshoot of their village and reorganized themselves into a group known as Dzuliekima Youth Organization. However, after several meetings, consultations and conducting awareness programmes about the benefits of conservation, the whole community agreed to the total ban on hunting and logging.

It was not just about the decision to ban logging and hunting, but rules and regulations were made and further steps were taken to enhance the practicability of their conservation effort. Since the implementation of total ban on logging for commercial purposes in 1996, to further reduce the felling of trees on a large scale, the villagers were allowed to cut down trees only for their basic necessities, such as for firewood and timber for house construction. And in order to prevent further deforestation, every household is permitted to cut down only a truck load of firewood in a year. Moreover, if any of the villagers needs timber for house construction they have to submit an estimate of the house and accordingly the Village Council allots them the number of trees for the construction. The Council also maintains a file where they keep a record of the names of individual, number of households to keep a track of those who have already been granted the permission to take timbers from the Village forest for house construction. They not only stopped cutting down the trees but started to plant more trees, especially the *Khel* youths, the Student Union and the younger age-group have started to plant trees on the side streets, in and around the village vicinity and on the road sides leading to their village.

Hunting as an activity was a good means of income for the villagers because there is a high demand for the wild games as it is considered a delicacy and is also utilized for its zoo-therapeutic property by the Nagas. Moreover, hunting was considered as a traditional sport by most of the Nagas, thus, it was an integral part of their life. All such factors combined with no measures to restrain the excessive hunting, led to rampant hunting which resulted in the rapid decrease of the wildlife population and many rare and endangered species were also killed in the process. Initially, the people were concerned about the large scale killing of Blyth Tragopan, an endangered pheasant which was found in plenty in Khonoma, so they decided to prohibit the hunting of the pheasant in their village and in 1998, the community established the Khonoma Nature Conservation and Tragopan Sanctuary which covered about 70 km^2 of the forest area. Later on, the community together with their leaders again held talks and meeting to stop the indiscriminate hunting. At the initial period, hunting was prohibited but it was allowed for 15 or 20 days during the hunting season (December-March). But it has its consequences as after almost months of conservation, everything was hunted down during the 15-20 days relaxation. So, the leaders along with the community held discussions about the problem and decided to completely ban hunting. However, there were oppositions because wild games were part of their diet, it was deeply connected to their culture and for some of the villagers hunting and selling the wild games was the only source of livelihood. But after many discussions and negotiations to find alternative means for those who were dependent on hunting, they came to a consensus and in the year 2000, total ban on hunting of wildlife in Khonoma village area was imposed by the Village Council.

They also made rules and regulations to penalize the offenders and the Khonoma Youth Organization (KYO) was entrusted to implement these rules and impose fines whenever necessary. According to the rules laid down, if anyone is caught hunting, whether alone or in groups, each person has to pay an amount of ₹ 3000 as penalty and their guns will be ceased. And out of the sum of ₹ 3000, ₹ 1000 is given to the reporter who reports about the illegal hunting, ₹ 1000 to the *Khel* Youth of the *Khel* in whose area the infiltration took place and ₹ 1000 is kept by the KYO. This is done so as to encourage people to report honestly and the youths to be stringent in regulating the hunting prohibition. Even children were not exempt, if they were seen with sling-shots or caught hunting with it, they were fined ₹ 100. "*This was done not as a punishment to them but to inculcate the importance of conservation and a sense of responsibility to care for the environment among the children from a young age*".

But an ordeal faced by many of the farmers was the destruction of the crops by the wild animals. In order to deal with this problem, they made a regulation to allow the farmers to kill the animals after they take permission from the KYO, but they were not allowed to sell the games killed in their field. In addition, in order to prevent the people from taking advantage of the permission allotted to them, they have to make a list of the animals that are destroying their crops and have to shoot only those animals that they have written on the list, if not, they have to pay the penalties. Also the Youth Organization permits them specific number of days according to the type and number of animals that are destroying the crops and within the given period they have to hunt down the animals.

The question of the practicability of imposing the total ban on hunting and how to enforce the regulation was raised. The village leaders, *Khel* Unions, Youths along with the rest of the community decided to make a register with a written pledge that "*every person and family gives their assurance not to hunt anymore and abide by the rules and regulations of the village and give their level best to protect their environment*". One register each was circulated in the three *Khels* and another one in Dziliekie which is an offshoot of the village, where the head of every household signed on behalf of himself and his family to abide by the pledge and to face any consequences if the rules were broken. And in 2011, on the Decennial Celebration of Successful Conservation, the registers were brought, and the Youth President of every *Khel* along with the President of Khonoma Youth Organization read out the agreement, made the pledge to conserve the forest and signed the declaration.

At the outset, the Khonoma Village Council supervised all the matters of ban on hunting and logging, but with many official duties and responsibilities to attend to, it were difficult for the members to keep a vigil on the activities and there was still instances of people carrying out illegal hunting and logging. So, the Village Council members and community held meeting and decided to give the responsibilities to monitor on the illegal activities to the youths because the youths were the main infiltrators and if they were given the responsibilities they themselves have to first stop from such activities in order to control the others who were illegally carrying out logging and hunting. As one of the respondents said, "*it is like the hunters being hunted*". Thus, the power to check and monitor the activities of illegal hunting and logging was handed over to the Khonoma Youth Organization. Further, to enhance their conservation effort, the people of Khonoma send official notice to the neighbouring villages, newspapers, various organizations and put up notice board about the prohibition on hunting and logging in Khonoma and the penalties that entailed if any individual commit any such offence within the vicinity of Khonoma village. Apart from ban on logging and hunting, the villagers of Khonoma also took up other initiatives to maintain their environment. Some of the initiatives are —

(a) **Ban on Collection of Forest Produces for Commercial Purpose:** Khonoma has plenty of wild vegetables, fruits, herbs, plants with medicinal and aromatic values and wild flowers, which were collected and sold by the women. But in order to conserve the flora of the village, the community decided to impose a ban on such commercial activities. The villagers are allowed to collect the wild leaves, vegetables, herbs etc. for their personal use and they are allowed to take it for their relatives and friends staying outside the village, but taking the forest produces outside for selling is strictly prohibited. The only wild product that is allowed to be sold commercially is the wild apple, for which Khonoma is well-known. The villagers mentioned that since the wild apples are found in plenty in the village, if they don't utilize it would just rot away and be wasted, so the villagers, mostly women and children collect it from the forest, preserve it in the form of pickle, jam, dry it with sugar and salt, make juice and sell it in the market. And for many of these women and children, it is a very good source of income for them. In addition, many of the villagers have started to preserve the wild plants and vegetables, medicinal plants and herbs by planting it in their own *jhum* fields and in their homesteads as well.

(b) **Prohibition on Use of Chemicals in the Fields and Rivers:** Organic farming is a practice which the people of Khonoma have been following since their ancestor's time. Khonoma is well known for their traditional agriculture system of Alder based *Jhum* cultivation and Terrace fields for wet paddy cultivation. The Alder tree (*Alnus nepalensis*) restores the fertility of the soil by fixing atmospheric nitrogen into the soil with its root nodules and it also enhances the crop production and reduces soil erosion. The people of Khonoma being aware of the advantages of the Alder tree have evolved a system of managing their *jhum* field by planting it in their fields and surroundings. This has helped them to sustainably cultivate the land without degrading the land and forest, and in addition, their firewood requirement also meet from the alder trees grown in their fields and this has greatly reduced the pressure on the forest and has helped in conserving the forest. Thus, it has contributed to the greener environment, ample supply of clean water, fertile soil and less soil erosion in Khonoma. Even at present, the community follow their traditional method of cultivation and use of chemical fertilizers and pesticides are discouraged. They also strictly prohibit the use of chemicals and explosives during fishing, instead roots of plants and creepers and traps are used to catch fishes.

(c) **Use of Solar Light:** In order to save energy, solar lamps were distributed to every household in the village. And instead of using electric lights, they have fixed solar bulbs on bamboo posts to light up the village streets.

(d) **Maintenance of Incinerator:** All the three *khels* in the village have an incinerator each where they collect their waste materials and burn it off.

(e) **Keeping Dustbins all-round the Village:** All-round the village surroundings, the community have kept dustbins made of bamboos and other eco-friendly materials. And every household in the village maintains dustbins to dispose the household waste. A good practise here is that even the children as young as 4-5 years do not simply throw away the sweet covers or papers here and there, but throw it in the dustbins and if they see any waste littering in the streets they pick it up and throw it in the nearby dustbins.

(f) **Sanitation Drive:** The Khonoma Student Union (KSU) has taken the initiative to clean the village surroundings at least once in a month. Thus, in the early hours of the selected day, all the students, starting from Class V to Class X, sets out to clean their surrounding, pick-up the litters and sweep the village street etc. Also the Khonoma Women were actively involved in creating awareness about sanitation in the village and took initiatives and gave support to all the activities to conserve the environment of the village.

Such Conservational Approach initiated by the community was appreciated and different organizations like Aaranyak Guwahati, Bombay Natural History Society, Centre for Environment Education Ahmedabad, Equations Bengaluru, Kalpavriksh Pune, Nature's Beckon, North-East India and Nagaland Empowerment of People through Economic Development extended their assistance to the people of Khonoma in their effort to conserve their environment. The Union Ministry of Tourism and Culture of India recognizing the potential of the Conservation Approach of the Khonoma community decided to promote eco-tourism in the village and in the year 2003, granted the villagers an amount of ₹ three crore for the Green Village Project through the Tourism Department of Nagaland. The fund was utilized by the villagers "to start a tourism initiative, to provide basic civic amenities and hygienic measures, reinforce community infrastructure and prepare the village to receive and showcase to visitors its past and its present" (Pathak and Hazarika, 2012: 87). The rich bio-diversity of the village, the flora and the fauna and the prominent topographical features of the village were documented as an undertaking of the Green Village Project. The Khonoma Tourism Development Board along with the villagers and with the help of a team of experts from various fields also carried out an Environmental Impact Assessment in order to have a proper understanding of the bearings that eco-tourism will have on the local population and the local environment and to strengthen the capacity of the local community in various sectors based on the assessment. The active participation of the community in the conservational effort along with the support of the NGOs and the government resulted in the productive outcome of the Green Village Project and in October 25, 2005, Khonoma was inaugurated as the first "Green Village" in India by the then Chief Minister of Nagaland, Mr. Neiphiu Rio. Thus, Khonoma – the Green Village is a model village for many communities in the country as well as outside India.

Challenges and Opportunities

The success of the conservational effort of the Khonoma community was not without any obstacles. It has its own set of problems and challenges, at the same time it opened up new avenues and opportunities for the community. Some of the challenges faced by the Khonoma community are —

(i) **Question of Livelihood:** For many of the villagers logging business, hunting, collecting and selling of wild products was the main source of income which they utilize for their children's education, medical bill and household expenses etc. Under such circumstances, they have to deliberate over the question of the alternative means of livelihood to which they could recourse to once all these activities were stopped.

(ii) **Opposition from within the Community:** As many of the villagers were dependent on hunting and logging, there was strong opposition against the total ban on logging and hunting. Especially many of the unemployed youths of Khonoma were engaged in the logging business, as they have no other means and logging was a good source to earn large amount of money easily and quickly, they were totally against the ban on logging. Some of them even went to the extent of leaving Khonoma and forming an organization in Dzuliekima, which was an off-shoot of Khonoma village. Such oppositions were a hindrance to their conservational effort.

(iii) **Conflicts:** The difference in opinion about the total ban on hunting and logging, the opposition against it lead to division among the people of Khonoma and "*it nearly destroyed the village*". Meeting, consultations of the leaders and community were disrupted by those who were against the banning of logging and hunting. During this time "*there were continuous heated arguments among the villagers everyday*". Moreover, there was intervention from the underground militants as they used to receive tax from the logging trade, which will be stopped once logging is banned.

(iv) **Controlling the Banned Activities:** Even after the community decided to ban logging and hunting, there were individuals still carrying out these activities illegally. There were also cases where even after they were caught red-handed, some of them were not willing to pay the fines. Such rebellious actions and controlling it was a challenge they have to deal with for their conservational effort to be successful.

In spite of all these difficulties, the people of Khonoma were successful in their endeavour to conserve their environment because of their strong community solidarity. Many of the individuals who were actively engaged in logging and hunting decided to give up their activities and support the community conservational effort because "*it was for the good of our people, for our village and for the future generations*". There are instances of people leaving the village in search of new means of livelihood "*instead of staying back and breaking the rules and bringing bad name to the village*". Here, the needs of the community, society and the village took precedence over their individual needs because "*since the time of our ancestors, we lived and shared together all our failures and success, and the unity and progress of our village is more important than our individual accomplishment*". Besides, the meetings and consultations were held to create awareness among the people about the advantages of the conservation and convince them to give up those activities that would result in the destruction of their environment. Moreover, the retrospection about their traditional practices which was based on nature and preservation of the natural resources, the practices and wisdom of their ancestors which made it possible for them to live and enjoy in their land now, was a strong base for them to resort back to the conservational activities. All these factors along with the determination of the community to conserve their village environment made it possible in the success of their conservational endeavour. And the success of their conservation led not only to a safer, cleaner and richer environment, it also opened up new avenues and opportunities for the people of Khonoma. Some of these were:

(i) **Green Village:** With the naming of Khonoma as the first Green Village in India, Khonoma became a tourist destination where tourists and researchers visited everyday throughout the year. This opened up new contacts with outsiders and many new ventures.

(ii) **New Economic Avenues:** Agriculture and government employment is not the only option for the Khonoma people any longer. Home stays to host the guests, youngsters working as tourist guide, plying taxis from Kohima to Khonoma, cultivation of cash crops, marketing their traditional products are some of the new economic activities through which they are earning their livelihoods.

(iii) **Improvement in Village Infrastructure:** Khonoma village was granted ₹ 3 crore for the Green Village Project and this amount was utilized by the villagers to construct roads, footpaths, culverts, waiting sheds, installing solar lights in the village streets and distributing solar lamps to the households etc. Thus, there was an over-all infrastructural improvement in Khonoma. It also provided employment to the villagers as all the development activities were carried out by the villagers themselves and the workers were paid from the project money.

(iv) **Rejuvenation of their Traditional Practices and Culture:** It also gave them an opportunity to revive their traditional practices of preserving their natural resources and using it sustainably. Documentation of their festivals, folklores and Traditional Ecological Knowledge, names of flora and fauna found in Khonoma and their usages was another benefit for the villagers. And it also gave them an opportunity to showcase their tradition and culture to the outside world.

Conclusion

The strong determination of the Khonoma community to conserve their environment, their courage to risk everything even the means of their livelihood for the good of the community and their environment, is a story to be upheld in mankind's history. Even now, it has its own challenges such as the struggle to ensure that the benefits reach every stakeholder as there are claims that only a few gets the benefits. The community are also struggling to find out ways and means to sustain their conservational initiatives. Here, the Government, NGOs and various organizations can give them logistic support, financial aid and support in various ways. However, in spite of all these challenges, the community initiative to safeguard their natural heritage should be lauded as the outcome of the environment conservational effort of the people of Khonoma was a better, greener, and richer environment for their village. In addition, it has consequential impact on their social and economic life. The success of their conservational effort itself highlights the strong community bond of their society. The debates, oppositions, discussions, resolutions with regard to the conservation issue also show the underlying social mechanism of rationally resolving conflicts among the community and the social cohesion that upholds the community in times of crises. It also brought about new economic avenues for the people. Khonoma village became a tourist attraction because of their green and clean environment, and their conservational activities and the villagers started Home Stays to host the guests from outside, the youths were employed as tourist guide and better earning opportunities for the taxi drivers as more tourists come to visit Khonoma. Moreover, it gave them the opportunity to revive many of their traditions and showcase their tradition and culture to the outside world. Also, the active involvement of the women in the conservational activities of the village gave them an opportunity to contribute to their community, at the same time it empowered them by uplifting their economic life and enhancing their social status. And the role played by the youth and the responsibilities given to them in monitoring the conservation of their environment placed them in a significant position where they have a crucial role to play and contribute for the community and for the future of the village. And most importantly, Khonoma entered the annals of the environment history for their conservation effort and earned the name – the first ever "Green Village" in India.

REFERENCES

Aier, Anungla and Changkija, Sapu (2003). "Indigenous Knowledge and Management of Natural Resources", in T.B. Subba and G.C. Ghosh (ed.), *The Anthropology of North-East India* (337-378), New Delhi: Orient Longman Private Limited.

Arnold, David (1996). *The Problem of Nature*, UK: Blackwell Publishers.

D'Souza, Alphonsus B. (1995). *Traditional System of Forest Conservation in North-East India: The Angami Tribe of Nagaland*, Guwahati: North-Eastern Social Research Centre.

Government of Nagaland (2004). *Nagaland State Human Development Report*, Kohima: Department of Planning and Coordination.

Government of Nagaland (2014). *Annual Administrative Report, 2013-14*. Kohima: Department of Forests, Ecology, Environment and Wildlife.

Hutton, J.H. (1921). *The Angami Nagas*, London: MacMillan & Co. Limited.

Khonoma Tourism Development Board (2004). *Environmental Impact Assessment Report*, Kohima: KhonomaTourism Development Board.

Longchar, A. Wati (1995). *The Traditional Tribal Worldview and Modernity*, Jorhat: N. Limala Lkr.

Nagaland Environmental Protection and Economic Development (NEPED) and International Institute of Rural Reconstruction (IIRR) (1999). *Building upon Traditional Agriculture in Nagaland, India*, Kohima: NEPED and IIRR.

Nshoga, A. (2008). *Traditional Naga Village System and Its Transformation*, Delhi: Akanksha Publishing House.

7

Plant Diversity in Community Managed Forests of Meghalaya

Aabid Hussain Mir and *Krishna Upadhaya*

Introduction

Forest resources have been considered as an integral part of life supporting system for several indigenous communities in many parts of the world. For these rural communities, the forest is a critical source of water, timber, wildlife, and identity. Forest resources lie at the heart of their social, political, and economic life as the primary source of livelihood, nutrition and employment (Tiwari *et al.*, 2013). The global indigenous population of approximately 300 million people is composed of about 5,000 distinct indigenous cultures worldwide, living in climatic conditions ranging from the Arctic Circle to the tropical rain forests. Of which nearly half, i.e. 150 million are living in Asia, about 30 million in Central and South America and a significant number of them are living in Australia, Europe, New Zealand, Africa, and Soviet Union. These indigenous communities live in close proximity of forests and have learnt to control and preserve the bio-diversity since long time (Mir and Upadhaya, 2015). Many areas inhabited by indigenous people coincide with some of the world's remaining major concentrations of bio-diversity. They view conservation areas as an integral and functional part of the landscapes in which they live. Although indigenous people make up only 4 per cent of the world's population, they occupy as much as 85 per cent of the world's protected areas (IUCN/WCPA, 2004). This convergence of bio-diversity-significant areas with indigenous territories presents an enormous opportunity to expand bio-diversity conservation efforts beyond national parks and reserves.

During last few decades, increase in population, urbanization and industrialization has led to increased use of forest resources, leading to a fast depletion of natural forest cover. The world forest cover has decreased from 4128 million ha in 1990 to 3999 million ha in 2015 (FAO, 2015). This represents a change from 31.6 per cent of global land area in 1990 to 30.6 per cent in 2015 (FAO, 2015). According to the World Conservation and Monitoring Centre, more than 8000 tree species are endangered worldwide. Conservation biologists estimate that about 25 per cent of the total life forms may become extinct during the next few decades (IUCN, 2003). Human activities like collection of timber, fuel-wood, non-wood forest products and grazing and trampling causes varying level of disturbance and alteration in the habitats of many forest species resulting in spatial and temporal variation in species richness, composition, productivity and regeneration status of plant species (Heywood and Watson, 1995; Kang *et al.*, 2007). Nationalization of forests also alienated the people from forests leading to massive deforestation (Schulte and Sah, 2000). While forest cover continues to shrink globally, strategies to revert this trend are often controversial (Geist and Lambin, 2002). Nonetheless, a general conformity exists that a mix of different forest conservation strategies are desirable which integrate public-private and community-managed areas (Bray *et al.*, 2008).

The State of Meghalaya in North-East India is a part of Indo-Myanmar hotspot. The state is rich in plant diversity and harbours about 3,128 flowering plants, of which 1,236 species are endemic (Khan *et al.*, 1997), 363 are rare and threatened (Upadhaya *et al.*, 2013) and 834 are used as traditional herbs (Mir *et al.*, 2014). About 90 per cent of the land belongs to the local people. More than 80 per cent of people live in the rural areas and are dependent on forests for their livelihood (Tiwari *et al.*, 2013). The local people have an age-old community forest management systems that are still functional and are contributing to conservation and sustainable utilization of forest resources (Tiwari *et al.*, 2010). The state has a total forest area of 15,657 km^2, of which 1,027.20 km^2 is under the control of State Forest Department, and constitutes only 4.58 per cent of the total geographical area of the State and 6.56 per cent of the total forest area of the State. Rest of the area (14,630 km^2) is either private or clan/community owned and is under the indirect control and management of the Autonomous District Councils (Forest and Environment Department, 2016). The forests are classified into different types depending on their intended use. According to Rodgers (1994) and Tiwari *et al.* (2010) traditionally managed forests consists of (a) Sacred groves (Law-Lyngdoh/Law-Kyntang/Law-Niam), (b) Reserved/restricted forests (Law-Adong), (c) Village forest (Law-ri-Sumar/Law-Shnong), and (d) Forest of a cluster of villages (Law-Raid). The traditional management practices vary in these forests. The permission to harvest the forest produce ranges from extraction of any forest produce including NTFPs and timber in case of village forests to moderate extraction (NTFP and fuel-wood) in case of restricted forest and there is strict prohibition on extraction of any kind of material from sacred groves.

Since last few decades, the forest resources of the state are undergoing rapid transformation due to urbanization, mining and extraction of forest products, besides continuance age-old practice of shifting cultivation. Due to these activities, many of the species that were considered as common have become rare and threatened in the state. Though, the importance of the sacred groves of the state have been emphasized by a number of workers (Khan *et al.*, 1997; Tiwari *et al.*, 1998; Upadhaya, 2002), but there is limited information on other types of community forests. Therefore, the current study was carried out to understand the floristic composition and structure of the forests under traditional management practice gradient. In addition to this, reasons for disappearance of such forests and effective strategies for conservations are also discussed.

Materials and Methods

Study Area

In order to study the effectiveness of forest management practices, four different types of community forests were analysed for woody species diversity (Table 7.1). The forests surveyed included a Sacred forest (Law Kyntang), Village restricted forest (Law Adong), Village forest (Law Shnong) and forest of a cluster of villages (Law-Raid), hereafter called as LK, LA, LS and LR respectively. All the patches are located in and around Cherrapunjee area of East Khasi Hill district of the state. The selection of forest at a single region was done in order to overcome the variations caused by varied topography and climatic conditions. The vegetation of the area can be classified as subtropical broad-leaved humid forest and represents the remnants of the climax forest of the area (Champion and Seth, 1968). Under the Koppen climate classification, the areas feature a "*subtropical highland climate*" with an extraordinarily rainy and lengthy monsoonal season. The wet period extends from April to October during which more than 80 per cent of the total rainfall occurs. The dry period extends from November to March with rainfall < 22 mm. The mean annual rainfall for the period 2008-2010 of the area was 11,873 mm and the average maximum and minimum temperature was 22° C and 14° C.

Table 7.1: Location of the Study Site and Current Human Disturbances

Parameter	*Law Kyntang (LK)*	*Law Adong (LA)*	*Law Shnong (LS)*	*Law Raid (LR)*
Latitude	25°14.913′	25°14.354′	25°14.323′	25°15.147′
Longitude	91°42.044′	91°43.841′	91°43.864′	91°42.698′
Elevation (m)	1181	1074	1115	1243
Management institution	Village Council	Village Council	Village Council	Group of village Council
Level of extraction	Low	Medium	High	High
Current disturbances	Fuel-wood harvesting, NTFPs collection and building roads.	Fuel-wood harvesting, NTFPs collection, grazing, building roads and fire.	Extraction of timber, fuel-wood harvesting, NTFPs collection, clearing forest land for agri-culture, grazing, building roads and fire.	Extraction of timber, fuel-wood harvesting, NTFPs collection, clearing forest land for agri-culture, grazing, building roads and fire.
Disturbance index	15	25	35	35

Source: Field Study.

Data Collection and Analysis

In each forest, a belt transect of 20 m × 250 m (0.50 ha), sub-divided into quadrats of 10 m × 10 m size was laid. All woody species of ≥ 5 cm diameter at breast height (dbh) were measured. The specimens were collected and identified with the help of Regional floras (Kanjilal *et al.*, 1934-1940; Balakrishnan, 1981-1983; Haridasan and Rao, 1985-1987) and Herbaria at BSI (Botanical Survey of India) Eastern circle. Community parameters such as frequency, density, dominance, importance value index and basal cover of the species was computed following Misra (1968), Mueller-Dombois and Ellenberg (1974). Species richness was determined directly by counting the number of species in each site. Shannon-Wiener index of diversity, Pielou's evenness index and Simpson dominance index were calculated according to Magurran (1988). All the identified individuals were categorized into different diameter classes, viz., 5–15, 16–25, 26–35, 36–45, 45–55, 56–65 and > 65 cm to analyse the population structure of the species.

The disturbance index for each site was computed following Mir *et al.* (2016). A score of five was assigned to each of the human disturbance factors, viz., extraction of timber, fuel-wood harvesting, NTFPs collection, clearing forest land for agriculture, grazing, building roads and fire. Any site with all these disturbances would have a total score of 35. Species, whose distributions are restricted to Meghalaya or to the North-Eastern region including Indo-Myanmar hotspot, were considered as endemic. Threatened category includes those plants that are classified as per the Red Data Book of Indian plants by Nayar and Sastry (1990), Walter and Gillett (1998), Ved *et al.* (2005) and IUCN (2015) Red List.

Results

Traditional Institutions and Forest Management

In Meghalaya, all three major ethnic tribal groups, namely, the Khasis, Jaintias and the Garos have their own traditional institutions, through which they are managing and conserving their forest resources (Tiwari *et al.*, 2013; Mir and Upadhaya, 2015; Sarma, 2015). For example in a typical Khasi community, the larger territory base (Hima) consisting of a number of villages is headed by 'Syiem'. The head of the village (Rangboh Shnong, Sordar or Myntri Shnong) is in charge for governance of forests at village level. Among Jaintias, the congregation of villages (Raids) are headed by 'Dollois', and in Garos, the head is called as 'Nokma', who are responsible for performing the executive, magisterial, religious and ceremonial functions. These traditional institutions have classified the forests into different categories, which help in conservation of plant species and in the order of protection they can be arranged as sacred groves > restricted forests > village forests > forests of cluster of villages.

Characterization of Anthropogenic Disturbance

The community forests are subject to a wide variety of disturbances including extraction of timber, fuel-wood harvesting, NTFPs collection, clearing forest land for agriculture, grazing, building roads and fire. The intensity of disturbance was highest in LR and LS (disturbance index = 35 each), followed by LA (DI = 25), whereas LK was least disturbed (DI = 15) (Table 7.1). At the site LR, maximum number of cut stumps were recorded (134, 25% individuals ha^{-1}), followed by LS (106, 11% individuals ha^{-1}), LA (58, 6% individuals ha^{-1}) and LK (30, 3% individuals ha^{-1}). The fuel-wood extraction was an important factor for forest disturbance in all the stands in comparison to timber extraction. Although small in magnitude, all the stands are threatened by road building and fuel-wood extraction, indicating that the disturbance is also spreading to forests under high protection. Grazing was common in the highly disturbed stands (LR and LS). Except LK, all other forests were affected by fires.

Floristic Composition and Diversity

During the current study, a total of 141 woody species (≥ 5 cm dbh), distributed in 56 families and 89 genera were recorded (Table 7.2). The species richness of the sites LA, LK, LS and LR were 77, 72, 66 and 53 respectively. The Shannon – Wiener diversity and Simpson dominance indices were 3.61 and 0.04, 3.55 and 0.05, 3.49 and 0.05, and 2.78 and 0.18 respectively for site LK, LA, LS and LR. The corresponding evenness indices value was 0.82, 0.84, 0.83 and 0.70 respectively (Table 7.2). Overall, Lauraceae with 7 genera and 22 species, Rubiaceae (5 genera, 7 species), Fagaceae (3 genera, 7 species), Theaceae and Rosaceae (3 genera, 6 species each), Aquifoliaceae (1 genera, 5 species), Fabaceae (4 genera, 4 species), Elaeocarpaceae (2 genera, 4 species), Oleaceae (3 genera, 4 species) and Symplocaceae and Myrtaceae (1 genera, 4 species each) were among the first ten dominant families. Seven families (Araliaceae, Clusiaceae, Ericaceae, Euphorbiaceae, Magnoliaceae, Moraceae and Rutaceae) were represented by 3 species each, nine families (Caprifoliaceae, Apocynaceae, Lamiaceae, Euphorbiaceae, Phyllanthaceae, Primulaceae, Proteaceae, Salicaceae and Schisandraceae) with 2 species each, whereas there were 29 families (Actinidiaceae, Anacardiaceae, Annonaceae, Arecaceae, Aristolochiaceae, Asteraceae, Berberidaceae, Betulaceae, Calophyllaceae, Celastraceae, Connaraceae, Cornaceae, Daphnephyllaceae, Ebenaceae, Erythroxylaceae, Juglandaceae, Lardizabalaceae, Malpighiaceae, Menispermaceae, Myricaceae, Myrsinaceae, Pandanaceae, Podocarpaceae, Aceraceae, Sapotaceae, Sterculiaceae, Styracaceae and Urticaceae), which were represented by only one species. Family, Lauraceae was dominating in all the sites with 6 genera and 11 species at LA, 6 genera, 9 species (LS), 6 genera, 12 species (LK) and 4 genera, 7 species at LR, respectively. The number of endemic and threatened species was low in village- and Rain-forest as compared to the sacred grove and restricted forest (Table 7.2).

Table 7.2: Species Diversity, Density and Basal Area in Different Types of Community Managed Forests in Meghalaya

Parameter	*LK*	*LA*	*LS*	*LR*
Species richness	72	77	66	53
Number of genera	54	55	52	40
Number of families	38	39	33	29
Number of threatened species	15	21	16	14
Number of endemic species	23	25	15	11
Basal area (m^2 ha)	58.18	38.93	21.06	8.77
Density (ha^{-1})	1026	964	992	546
Shannon Wiener index	3.61	3.55	3.49	2.78
Simpson dominance index	0.04	0.05	0.05	0.18
Pielou's evenness index	0.84	0.82	0.83	0.70

Note: LK = Law Kyntang, LA = Law Adong, LS = Law Shnong and LR = Law Raid.
Source: Field Study.

Stand Density, Basal Area and Dominance

The density of woody species ranged from 546 (LR) to 1026 (LK) individuals ha^{-1} with a mean value of 882 ± 225 individuals ha^{-1} (Table 7.2). *Echinocarpus murex* was the dominant species (106 individual's ha^{-1}) followed by *Castanopsis kurzii* (100 individual's ha^{-1}) at LA. At LK, *Castanopsis tribuloides* (126 individuals ha^{-1}) and *Elaeocarpus acuminatus* (62 individual's ha^{-1}), and at LS, *Syzygium cuneatum* (148 individuals ha^{-1}) and *Ligustrum robustum* (76 individual's ha^{-1}), whereas in LR, *Pandanus odoratissimus* (220 individual's ha^{-1}) and *Ligustrum robustum* (42 individuals ha^{-1}) were the dominant and co-dominant species. The basal area ranged from 8.77 (LR) to 58.18 (LK) m^2 ha, with a mean value of 31.73 ± 21.5 m^2 ha (Table 7.2). In terms of basal area, *Castanopsis kurzii* was the dominant species (17.63 m^2 ha) followed by *Echinocarpus murex* (4.89 m^2 ha) at LA. In LK *Castanopsis tribuloides* (23.47 m^2 ha) and *Elaeocarpus acuminatus* (7.48 m^2 ha), and at LS, *Syzygium cuneatum* (3.98 m^2 ha) and *Syzygium tetragonum* (2.12 m^2 ha), whereas in LR, *Pandanus odoratissimus* (2.12 m^2 ha) and *Ligustrum robustum* (1.08 m^2 ha) were the dominant and co-dominant species (Table 7.3). In terms of dominance (IVI),

Castanopsis kurzii (IVI = 63.1) and *Echinocarpus murex* (31.5) in LS, *Castanopsis tribuloides* (61.5) and *Elaeocarpus acuminatus* (23.9) in LK, *Syzygium cuneatum* (40.7) and *Ligustrum robustum* (19.8) in LS and *Pandanus odoratissimus* (83.3) and *Ligustrum robustum* (27.9) in LR were the dominant and co-dominant tree species respectively (Table 7.3).

Population Structure

Distribution of species richness and density was high in 5-15 cm dbh class, which sharply declined with the increase in the girth class (Fig. 7.1). The density-diameter distribution yielded reverses J-shaped curves in most of the stands. However, in LK and LA, the stem density in > 45 cm dbh class was higher than the other sites (Fig. 7.1).

Threat to Community Forests

All the forest categories under study were under severe threat. Sacred groves are the most threatened due to a number of reasons such as erosion of traditional belief and practices, encroachment for agriculture land, human habitation and illegal extraction etc. (Upadhaya, 2016). Survey in different parts of the state reveals that many sacred groves have been converted to Law Adong and subsequently to Village forest and overexploitation of it has led to the development of a barren land. Similarly, overexploitation of other community managed forests such as Law Adong and Law Shnong is also forcing the local communities to encroach upon the sacred groves and is a common phenomenon in many parts of the state. The village people in rural areas living near the forests are poor and depend on these forests to meet their vital domestic necessities, such as fuelwood, wild-vegetables, fruits, medicinal plants etc. Thus, overexploitation of village forest has resulted in the depletion of resources; leading to encroachment of the sacred groves and posing a direct threat. Another threat that is of recent origin is the conversion of Community owned forest to Community Reserves (Upadhaya and Tiwari, 2015) in the state. These protected area categories are created as per the provisions of Section 36C of the Wildlife Protection Act, 1972 (as amended up to 2006). Though it is not clear on how this change in management would affect or improve the livelihoods of forest dependent people of the state, but it would have an adverse effect on the use of resources.

Table 7.3: Density (individual's ha^{-1}), Basal Area (m^2 ha) and Importance Value Index (IVI) of Dominant Tree Species in Different Types of Community Managed Forests in Meghalaya

Species	*Family*	*LK*			*LA*			*LS*			*LR*		
		D	*BA*	*IVI*	*D*	*BA*	*IVI*	*D*	*BA*	*IVI*	*D*	*BA*	*IVI*
(1)	*(2)*	*(3)*	*(4)*	*(5)*	*(6)*	*(7)*	*(8)*	*(9)*	*(10)*	*(11)*	*(12)*	*(13)*	*(14)*
Acer laevigatum Wall	Aceraceae	—	—	—	—	—	—	—	—	—	—	0.48	8.36
Acronychia pedunculata (Linn) Miq.	Rutaceae	—	—	—	—	1.08	—	—	—	—	—	—	—
Calophyllum polyanthum Choisy	Calophyllaceae	52	—	12.46	34	0.74	9.12	—	—	—	—	—	—
Camellia cauduca Cl. ex Brandis	Thaeceae	—	—	—	—	—	—	62	0.65	15.94	—	—	—
Castanopsis armata Spach.	Fagaceae	—	—	—	—	—	—	—	—	—	18	0.92	17.47
Castanopsis kurzii (Hance) Biswas	Fagaceae	—	—	—	100	17.63	63.08	—	—	—	—	0.20	—
Castanopsis tribuloides (Sm.) DC.	Fagaceae	126	23.47	61.53	—	—	—	—	—	—	10	0.37	8.44
Cinnamomum tamala Fr. Nees	Lauraceae	48	2.11	12.75	—	—	—	—	—	—	—	—	—
Diospyros kaki Thunb.	Ebenaceae	—	—	—	—	—	—	—	0.77	—	—	—	—
Drymycarpus racemosus (Roxb.) Hk.f.	Anacardiaceae	—	1.46	—	—	0.83	—	—	—	—	—	—	—
Echinocarpus murex Benth.	Elaeocarpaceae	—	—	—	106	4.89	31.53	—	—	—	—	—	—
Elaeocarpus acuminatus Wall. ex Mast.	Elaeocarpaceae	62	7.48	23.87	—	—	—	—	—	—	—	—	—
Elaeocarpus prunifolius (C. Muell.) Mast.	Elaeocarpaceae	—	3.98	13.42	—	—	—	—	—	—	—	—	—
Ficus nerifolia J.E. Sm.	Moraceae	—	—	—	—	—	—	—	—	—	18	0.28	11.34
Glochidion thomsonii Hk.f.	Euphorbiaceae	—	—	—	—	—	—	34	—	8.79	—	—	—
Helicia excelsa Bl.	Proteaceae	—	1.29	—	—	—	—	—	—	—	—	—	—

(*Contd.*)

Table 7.3 (*Contd.*)

(1)	*(2)*	*(3)*	*(4)*	*(5)*	*(6)*	*(7)*	*(8)*	*(9)*	*(10)*	*(11)*	*(12)*	*(13)*	*(14)*
Helicia nilagirica Bedd.	Proteaceae	38	—	8.08	36	0.65	9.69	—	—	—	—	—	—
Ilex excelsa Hk.f.	Aquifoliaceae	—	—	—	—	—	—	54	1.32	16.83	—	—	—
Ligustrum robustum (Roxb.) Bl.	Oleaceae	—	—	—	—	—	—	76	1.23	19.80	42	1.08	27.93
Lithocarpus dealbatus (Hk.f. & Th. ex Miq.) Rehder.	Fagaceae	—	2.08	8.08	40	2.83	15.41	—	—	—	—	—	—
Lithocarpus elegans (Bl.) Hatus ex. Soep	Fagaceae	50	2.72	13.75	—	—	—	38	0.81	12.47	—	—	—
Litsea elongata (Nees) Hk.f.	Lauraceae	—	—	—	—	—	—	64	1.17	18.93	10	—	—
Litsea laeta Wall. ex Nees	Lauraceae	—	—	—	—	—	—	—	—	—	16	—	8.03
Litsea salicifolia (Roxb. ex Nees) Hk.f.	Lauraceae	—	—	—	32	0.68	—	—	—	—	—	—	—
Myrica esculenta Buch. Ham. ex D. Don	Myricaceae	—	—	—	—	—	—	—	0.65	—	—	—	—
Pandanus odoratissimus (Lamk.) Linn.	Pandanaceae	36	—	—	72	1.54	17.69	—	—	—	220	2.12	83.26
Psychotria adenophylla Wall.	Rubiaceae	48	—	9.61	—	—	—	—	—	—	—	—	—
Randia griffithii Hk.f.	Rubiaceae	—	—	—	—	—	—	46	—	11.84	10	0.47	10.20
Randia wallichii Hk.f.	Rubiaceae	—	—	—	44	—	9.52	—	—	—	—	—	—
Sarcosperma griffithii Cl.	Sapotaceae	—	1.20	—	—	—	8.28	—	—	—	—	—	—
Styrax serrulatum Cl.	Styracaceae	56	—	11.69	—	—	—	—	—	—	—	—	—
Symplocos spicata Roxb.	Symplocaceae	—	—	—	—	—	—	46	—	10.38	—	—	—
Syzygium macrocarpum (Roxb.) Balak	Myrtaceae	—	—	—	—	—	—	—	—	—	22	0.32	12.47
Syzygium tetragonum (Wt.) Kurz.	Myrtaceae	32	1.15	—	70	0.84	15.11	—	2.12	14.09	22	0.92	19.32
Syzygium cuneatum (Duthie) Balak	Myrtaceae	—	—	—	40	—	9.03	148	3.98	40.73	—	—	—
Viburnum simonsii Hk.f. & Th.	Caprifoliaceae	—	—	—	—	—	—	22	0.58	—	—	—	—
Other species 62/66/56/43	29/31/23/22	478	11.2	124.7	390	21.3	111.6	402	7.78	130.2	158	1.17	297.3

Source: Field Study.

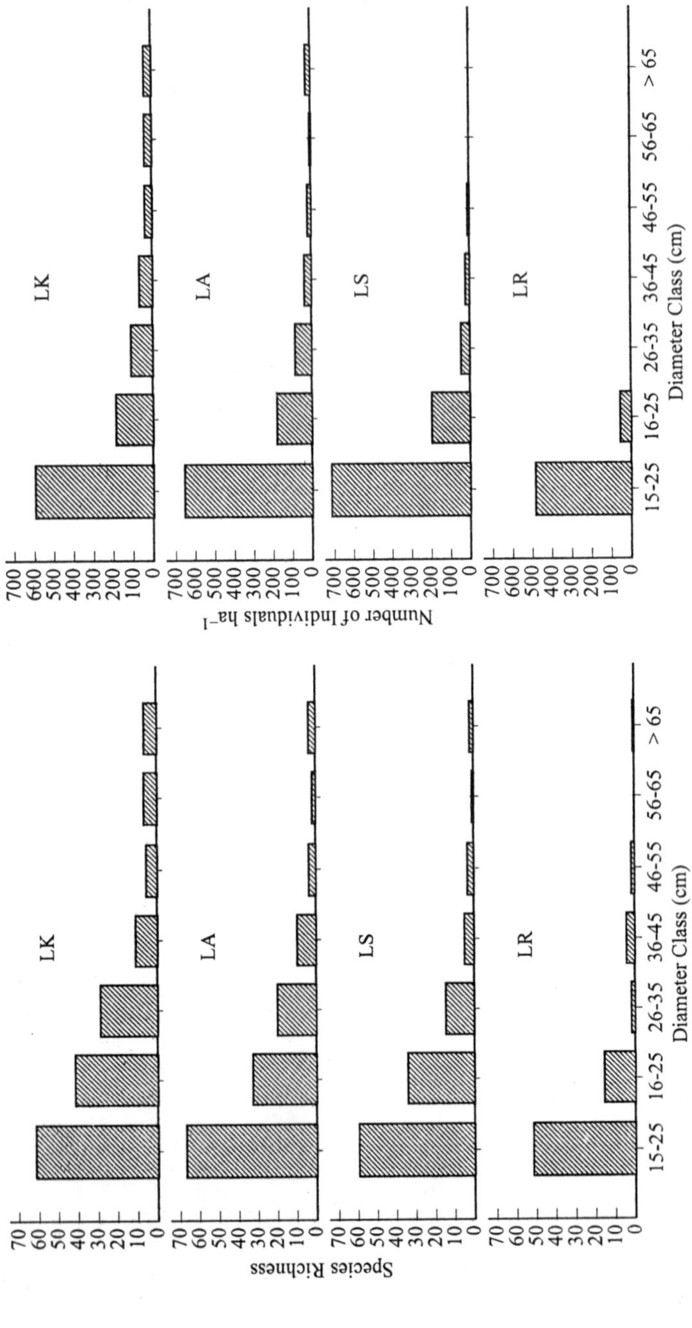

Figure 7.1: Species Richness and Density in Different Diameter Classes in Different Types of Community Managed Forests in Meghalaya.

Source: Field Study.

Discussion

Community-managed forests are rich in diversity and harbours many important plant species. These forests are viewed as one of the most effective platforms for conservation (Ormsby and Bhagwat, 2010). Species richness was found to be comparatively higher in forests with higher degree of protection/restrictions and less disturbed (LK and LA) than in highly disturbed and least protected forests (LS and LR). High species richness in LA as compared to LK could be due to the fact that intermediate levels of disturbance favours competitive species and those that tolerate disturbance to co-exist. Mild disturbance in the forests provides greater opportunity for species turnover, colonization and persistence (Upadhaya *et al.*, 2008). The low species richness in cases of LS and LR is due to repeated disturbance caused due to product extraction. Besides, the periphery of the forest is burnt during winter to promote herbage growth in the ensuing rainy season. Fire is a major barrier to natural forest regeneration because it decreases species diversity. It also causes a decline in the number of species recruiting from both the seed bank and resprouting, leading to an improvised community as documented at many subtropical sites (Upadhaya *et al.*, 2008). The species richness declined rendering the adverse effect of disturbances to all the compositional elements of the community. Frequent but low intensity disturbance (e.g., grazing and browsing, firewood extraction) may involve the combined effect of multiple factors, and may strongly affect forest structure and the ability of species to regenerate in the disturbed area (Kappelle *et al.*, 1996).

Several threatened and endemic species of North-East India such as *Alstonia scholaris, Melodinus monogynus, Ilex embelioides, Ilex venulosa, Aristolochia saccata, Carpinus viminea, Daphniphyllum himalayense, Elaeocarpus prunifolius, Engelhardtia spicata, Beilschmiedia assamica, Litsea laeta, Manglietia caveana, Michelia punduana, Fraxinus floribunda, Photinia cuspidata, Illicium griffithii* and *Schima khasiana* were mostly present in the forests with higher protection as compared to less protected forests. Agriculture, illegal logging and mining, are posing threat to such species. Even the forests with lesser protection are in constant threat of extinction. The site LR is affected because of mining activity by nearby cement factory. Building roads through forests fragments, endangers not only species habitat, but also makes it easier for illegal loggers to exploit the forest.

Despite the study site being located in a very high rainfall area with rugged terrain, the varied diversity status may be attributed to the different management practices which regulate human activities in these forests. The difference in type of management and human interference between each studied forest has considerable influence on community parameters of the forests, as evident from decrease in the species richness, density, basal area and diversity and dominance indices in the forests having high protection (LK and LA) towards less protected patches (LS and LR). This is also supported by the fact that the forests under open accessibility (LR and LS) have higher disturbance index than that of moderate accessibility (LA) to one with lowest accessibility (LK). These findings are in conformity with those of Chittibabu and Parthasarathy (2000), who showed reduced species richness, density, and basal area, as well as altered species composition in disturbed plots, as compared to the undisturbed plots. Anthropogenic factors such as mining, timber extraction, road construction and livestock grazing resulted in a sharp decline in forest vegetation attributes (Upadhaya *et al.*, 2013). Owing to the higher disturbances in less protected forests, most of the endemic and threatened species are taking refuge into the sacred and restricted forests.

Higher density and basal area of tree species in LK and LA than in LS and LR was primarily due to restricted extraction of timber and firewood from these forests. A steep decline in basal area and density in LR is because of greater human disturbances as people from more than one village extracts the forest produce from there. The decrease in tree density with increasing girth class (> 55 cm dbh) especially in LS and LR has revealed that mature trees have been extracted. The higher basal area in LK was due to the lesser human influence/disturbances and the presence of mature tree species. The reduction of basal area in other stands might be due to high disturbance and selective cutting of large and medium sized trees.

It may be concluded that traditional institutions in Meghalaya are the key mechanism in guarding the region's immense bio-diversity. These institutions manage the forests for variety of reasons, including resource enhancement and/or maintenance, countering ecological threat, expressing religious sentiments, cultural concerns and/or continuing traditional systems and managing bio-diversity concerns. This management system has not only helped in conserving the plant diversity, but is also ensuring its sustainable use by the local people. But owing to increased disturbances, more effective conservation strategies need to be undertaken. Forest fires, illegal logging, agricultural expansion needs to be checked and monitored. Along with this, the multi-stakeholder component of the management needs to be monitored and interactions between stakeholders and scientific communities should be maintained so as to develop effective conservation strategies, without disturbing livelihood patterns of the forest dependent people. The most critical challenge is to link management institutions for conserving forests with financial sustainability. Forest dependent people should be provided with alternatives for fuel-wood so as to lessen the pressure to these forests. Regular monitoring of plant diversity, soil health and modern scientific techniques also needs to be incorporated. The need of the hour is to make people aware about the effects of massive deforestation, shifting cultivation and other methods which are harmful for the vigour of the forest.

REFERENCES

Balakrishnan, N.P. (1981-1983). *Flora of Jowai*, 2 Vols. Botanical Survey of India, Howrah, India.

Bray, D.B., Duran, E., Romas, V.H., Mas, J.F., Velazquez, A., McNab, R., Barry, B.D. and Radachowsky, J. (2008). "Tropical Deforestation, Community Forests, and Protected Areas in the Maya Forest", *Ecology and Society*, 13(2): 56.

Chittibabu, C.V. and Parthasarathy, N. (2000). "Attenuated Tree Species Diversity in Human-Impacted Tropical Evergreen Forest Sites at Kolli Hills, Eastern Ghats, India", *Biodiversity and Conservation*, 9: 1493-1519.

FAO (2015). *Global Forest Resources Assessment: How are the World's Forests Changing?*, Food and Agriculture Organization of the United Nations, Rome.

Forest and Environment Department (2016). *General Description of Meghalaya*, Forest and Environment Department, Government of Meghalaya. *http://megforest.gov.in* (downloaded on 2/3/16).

Geist, H. and Lambin, E. (2001). *What Drives Tropical Deforestation? A Meta-Analysis of Proximate and Underlying Causes of Deforestation Based on Subnational Case Study Evidence*, Land-Use and Land-Cover Change (LUCC) Project, International Geosphere-Biosphere Programme (IGBP), LUCC Report Series, 4.

Haridasan, K. and Rao, R.R. (1985-1987). *Forest Flora of Meghalaya*, Vols. I and II, Bishen Singh Mahendra Pal Singh, Dehra Dun, Uttarakhand.

Heywood, V.H. and Waston, R.T. (1995). *Global Biodiversity Assessment*, UNEP, Cambridge University Press, Cambridge.

IUCN (2003). *Guidelines for Application of IUCN Red List Criteria at Regional Levels*, Version 3.0, IUCN Species Survival Commission, IUCN, Gland, Switzerland and Cambridge.

IUCN (2015). *The IUCN Red List of Threatened Species*, Version 2015-4. *www.iucnredlist.org*.

IUCN/WCPA (2004). *Indigenous and Local Communities and Protected Areas: Towards Equity and Enhanced Conservation*, By Borrini-Feyerabend, G., Kothari, A. and Oviedo, G., Best Practice Protected Area Guidelines Series No. 11, p. 139.

Kang, H.J., Chen, Z.L., Liu, P., Hao, C.Y. and Wei, F.M. (2007). "The Population Structure and Distribution Pattern of *Emmenopterys Henryi* in Dapanshan Natural Reserve of Zhejiang Province", *Acta Ecologica Sinica*, 27: 389-396.

Kanjilal, V.N., Kanjilal, P.C., Das, A., De, R.N., Bor, N.L. (1934-1940). *Flora of Assam*, 5 Vols. Government Press, Shillong, Meghalaya.

Kappelle, M., Geuze, T., Leal, M.E. and Cleef, A.M. (1996). "Successional Age and Forest Structure in a Costa Rican Upper Montane Quercus Forest", *Journal of Tropical Ecology*, 12: 681-698.

Khan, M.L., Menon, S. and Bawa, K.S. (1997). "Effectiveness of the Protected Area Network in Biodiversity Conservation: A Case Study of Meghalaya State", *Biodiversity and Conservation*, 6: 853-868.

Magurran, A.F. (1988). *Ecological Diversity and Its Measurement*, Princeton University Press, Princeton, New Jersey.

Mir, A.H. and Upadhaya, K. (2015). "Community Based Forest Management in India", in Upadhaya, K. (ed.), *Biodiversity and Environmental Conservation*, pp. 167-184.

Mir, A.H., Iralu, V., Pao, N.T., Chaudhury, G., Khonglah, C.G., Chaudhary, K.L., Tiwari, B.K. and Upadhaya, K. (2016). "*Magnolia Lanuginosa* (Wall.) Figlar & Noot in West Khasi Hills of Meghalaya, North-Eastern India: Re-collection and Implications for Conservation", *Journal of Threatened Taxa*, 8(1): 8398-8402.

Mir, A.H., Upadhaya, K. and Choudhury, H. (2014). "Diversity of Endemic and Threatened Ethnomedicinal Plant Species in Meghalaya, North-East India", *International Journal of Environmental Sciences*, 3(12): 64-78.

Misra, R. (1968). *Ecology Work Book*, Oxford & IBH Publishing Co., Kolkata.

Muller-Dombois, D.D. and Ellenberg, H. (1974). *Aims and Methods of Vegetation Analysis*, John Wiley & Sons, New York, USA.

Nayar, M.P. and Sastry, A.R.K. (1990). *Red Data Book of Indian Plants*, Volume III, Botanical Survey of India, Kolkata, West Bengal.

Ormsby, A.A. and Bhagwat, S.A. (2010). "Sacred Forests of India: Strong Tradition of Community-Based Natural Resource Management", *Environmental Conservation*, 37: 320-326.

Rodgers, W.A. (1994). "The Sacred Groves of Meghalaya", *Man in India*, 74: 339-348.

Sarma, K. (2015). "Land Ownership System and Status of Forests in Garo and Khasi Hills Autonomous District Councils of Meghalaya", in Upadhaya, K. (ed.), *Biodiversity and Environmental Conservation*, pp. 167-184.

Schulte, A. and Sah, S. (2000). "Historic Shift Towards Silviculture by People in Asia: A Review and Country Case Studies on Community Based Forest Management from Nepal, the Philippines and Indonesia", *Die Bodenkultur*, 51: 291-298.

Tiwari, B.K., Barik, S.K. and Tripathi, R.S. (1998). "Biodiversity Value, Status, and Strategies for Conservation of Sacred Groves of Meghalaya, India", *Ecosystem Health*, 4: 20-32.

Tiwari, B.K., Tynsong, H. and Lynser, M.B. (2010). "Forest Management Practices of the Tribal People of Meghalaya, North-East India", *Journal of Tropical Forest Science*, 22: 329-342.

Tiwari, B.K., Tynsong, H., Lynrah, M.M., Lapasam, E., Deb, S. and Sharma, D. (2013). "Institutional Arrangement and Typology of Community Forests of Meghalaya, Mizoram and Nagaland of North-East India", *Journal of Forestry Research*, 24(1): 179-186.

Upadhaya, K. (2002). Studies on Plant Biodiversity and Ecosystem Function in Sacred Groves of Meghalaya, Ph.D. Thesis, North-Eastern Hill University, Shillong, Meghalaya.

Upadhaya, K. (2016). "Ecological Analysis of Sacred Groves and Its Conservation", in Upadhaya, K. (ed.), *Biodiversity and Environmental Conservation*, pp. 167-184.

Upadhaya, K. and Tiwari, B.K. (2015). "Community Reserves: Challenges and Its Significance", in Kumar, N. (ed.), *Forest and Environment in India, Law, History, Culture and Modern Approaches*, pp. 213-223.

Upadhaya, K., Barik, S.K., Pandey, H.N. and Tripathi, O.P. (2008). "Response of Woody Species to Anthropogenic Disturbances in Sacred Forests of North-East India", *International Journal of Ecology and Environmental Sciences*, 34: 245-257.

Upadhaya, K., Thapa, N., Lakadong, J.N., Barik, S.K. and Sarma, K. (2013). "Priority Areas for Conservation in North-East India: A Case Study in Meghalaya-Based on Plant Species Diversity and Endemism", *International Journal of Ecology and Environmental Sciences*, 39(2): 125-136.

Ved, D.K., Kinhal, G.A., Ravikumar, K., Shankar, V.R. and Haridasan, K. (2005). "Conservation Assessment and Management Prioritization (CAMP) for the Wild Medicinal Plants of North-East India", *Medicinal Plant Conservation*, 11: 40-44.

Walter, K.S. and Gillett, H.J. (1998). *IUCN Red List of Threatened Plants*, Complied by the World Conservation Union, Gland, Switzerland and Cambridge.

8

Village Forest Development Committee (VFDC) in Forest Management in Champhai Conservancy in Town of Mizoram

Lalmalsawmi Hlondo

Introduction

Forest management can be explained as that branch of forestry whose function is the organization of a forest property for management and maintenance, by ordering in time and place the various operations necessary for the conservation, protection and improvement of the forest on the one hand, and the controlled harvesting of the forest on the other (Ram Prakash, 2006: 3). The *Dictionary of Forestry* defines Forest Management as, "The practical application of biological, physical, quantitative, managerial, economic, social, and policy principles to the regeneration, management, utilization, and conservation of forests to meet specified goals and objectives while maintaining the productivity of the forest" (Helms, 1998: 71). N.V. Brasnett "the primary object of good management is the provision of the maximum benefit to the greatest number of people for all times" (Brasnett, 1953: 9).

Joint Forest Management (JFM) in India

JFM is a forest management strategy under which the government (represented by the Forest Department) and the village community enter into an agreement to jointly protect and manage forest lands adjoining villages and to share responsibilities and benefits (Government of India, 2002). The Seventh Five Year Plan marked the official beginning of Joint Forest Management (JFM) in India as it was in June, 1990 that the

Government of India issued notification for rehabilitation of degraded forests with the participation of the local people in and around forest areas.

In the Ninth Five Year Plan period, in order to institutionalize people's participation, JFM was implemented through a pilot scheme called Samanvit Gram Vanikaran Samriddhi Yojana (SGVSY) in the country during 2000-2001, through two-tier decentralized institutional setup, namely: Forest Development Agencies (FDAs) in all Territorial/ Wildlife forest division level and Joint Forest Management Committees (JFMCs) at village level. The SGVSY scheme after gaining success was scaled-up to national level by the Tenth Five Year Plan period under the title National Afforestation Programme (NAP) operated by National Afforestation and Eco-Development Board (NAEDB) under the Ministry of Environment and Forests (MoEF, 2007).

Later as per the guideline of 2009 issued by the Central Government, the two-tier setup was further decentralized to a three-tier setup. State Forest Development Agencies (SFDAs) were added to operate above the two already existing bodies, which is to be constituted at the State level. SFDAs are registered under the Societies Registration Act, 1860, and will function as a federation of Forest Development Agencies (FDAs) and Joint Forest Management (JFM) is the central and integral part of all projects under the scheme (Principal Chief Conservator of Forests, Mizoram, 2009).

In Mizoram Joint Forest Management (JFM) was adopted in 18th of September, 1998, following the instruction of the Central Government's Notification of 1990. Under the National Afforestation Programme, Forest Development Agencies (FDA) and Joint Forest Management Committees (JFMCs)/Village Forest Development Committees (VFDCs) were constituted in Mizoram by the year 2002-2003. In Champhai, which is the area of the present study, Forest Development Agencies (FDAs) and Village Forest Development Committees (VFDCs) were started in the year 2003 (Department of Environment and Forest, Government of Mizoram, 2008).

Area of the Study

Champhai is the administrative headquarter of the district and it falls under the Central Forest Circle of Mizoram and is covered under one Forest Development Agency (FDA) namely Champhai FDA. Champhai Forest Division is further divided into 8 Ranges namely: Ngopa range,

Kawkulh range, Khawzawl range, Khawhai range, Farlawn range, Hnahlan range, Khawbung range and Champhai range. Champhai FDA has a total of 74 VFDCs operating in these 8 ranges and these ranges collectively have raised a total plantation area of 4,320 hectare (Department of Environment and Forest, Government of Mizoram, 2006).

Three localities in Champhai town/range were selected using simple random sampling technique. These localities are forest fringe localities located in the surrounding area of Champhai town within Champhai forest range, namely Kanan, Dinthar and Vengthar, whose inhabitants are basically manual workers dependent on forest for their essential needs.

Objectives of the Study

The present chapter is an attempt to highlight the challenges faced by VFDC in regeneration, management, conservation and protection of forest land covered under JFM in Champhai town.

Methodology

Descriptive survey method was employed for the present study.

Sampling

In Champhai town the total number of VFDC beneficiaries is approximately 500 households. A sample size of 60 respondents was selected from three localities, *viz.*, Kanan, Dinthar and Vengthar through simple random sampling (20 from each of the 3 localities).

Sources of Information

(i) *Secondary Sources:* articles, government documents, reports, published and unpublished books, internet sources etc.
(ii) *Primary Sources:* data was collected through field observation, interview and questionnaire.

Results and Findings

VFDC in the three localities was implemented by the Forest Department as per the guidelines of National Afforestation Programme (NAP). General meetings were held in each and every locality/village, in which the forest official brief the residents the aims, objectives, the need for creating such committee and the developmental schemes VFDC brings along with it towards the forests and also towards the locality/village. A memorandum of understanding was signed between the localities and the Forest department, when majority of the residents present in the meeting voted in favour of forming such a committee. In the same meeting, Managing Committee (MC)/Executive Body (EB) were formed, whose members were elected from among the general members. The function of the MC/EB is to represent the general members and look after the different activities of the VFDC and they form the core in the structure of the VFDC.

One of the major functions of the Managing Committee is to hold meetings which can be divided into three categories:

(1) *Entry Point Activity (EPA) Meeting:* In this meeting, the agenda is concerned with how, where, in what way, for what purpose, should the fund gained be utilized for the infrastructural development of the locality/village.

(2) *Work Meeting:* This meeting includes an agenda on the plantation work and the labour involved. Depending on the amount of labour required the committee makes the decision whereby usually the lesser privileged members/social target group of the locality are considered, and are given daily wage.

(3) *Awareness Meeting:* In the awareness meeting, the members of the Managing Committee/Executive body along with the forest officials, generally give awareness to the village people/ general body on importance of conservation, protection and regeneration of forest cover, maintenance facilities which members can avail/enjoy through EPA. This is the only meeting in which the General Body participates and it is regarded as a meeting-cum-training for the general members and is not conducted very often.

VFDC Plantation Profile of the Three Localities

In terms of the plantation projects, VFDC of the three localities together have raised a total plantation area of 240 hectare out of which 60 hectare has been successfully developed into plantation projects under different models of plantation. The plantation models taken-up by the three localities are:

(a) The VFDC of Kanan has adopted Aided Natural Regeneration (A.N.R.) as its model for its plantation project;

(b) The VFDC of Dinthar has adopted Artificial Regeneration (A.R.) as its model for plantation project; and

(c) The VFDC of Vengthar has adopted Bamboo plantation as its model for plantation.

Work Schedule and Maintenance

The work schedule is set by the Department of Forest and the above three steps are performed by the VFDC under the supervision of the member secretary who is an outfield personnel appointed by the Department, and the fund is incurred by the department. In the plantation work members of the VFDC basically provide labour force and they are paid wages, and as have been explained in the above paragraphs social target groups are given special preference from among the members.

The maintenance fund of the plantation is incurred by the Government, but this maintenance fund is provided only for three years, till the third year from the creation of the plantation and after the third year the initiative of maintaining the plantation is in the hands of the VFDC until the time when the plantations have grown fully to be harvested. When the plantations are harvested, the benefit is shared between the FD and the VFDC in the percentage of 50-50, and the 50 per cent share of the VFDC is further divided to 30 per cent to be shared among the beneficiaries or members and 20 per cent will be deposited for the developmental purpose of the VFDC. But according to the findings of the study, benefit derived from the harvest of plantations has not yet realised in the three localities and the reason for this was mainly because the plantation were still young and were at a developing stage.

With regards to harvest of other forest produce in the plantation such as Non-Timber Forest Products (NTFPs), the collection of NTFPs is permitted in the plantation site from the first year of its creation, whereas collection of fodder and firewood is permitted only from the 5th and the 6th year.

The findings of the study also highlight the major source of funds for the VFDC is Entry Point Activity (EPA) which is basically used towards the development of the localities/villages. EPA fund is gained by the VFDC when land is surrendered for creation of plantation projects and it is allotted only once at the beginning when plantation project is taken up by the VFDC. The amount of fund allotted through EPA depends on the amount of land surrendered for the purpose of creating plantation. The amount of EPA fund was ₹ 4000 per hectare initially, but was reduced to ₹ 1000 per hectare at the later stage of the scheme.

Knowledge Component of the Beneficiaries

Data was collected on the Knowledge component of the respondents about the working principles of the VFDC. Various options such as objective, functions, benefits, management plan, disposal of forest produce and role of different department and officials were provided. It was found through the data that 56 per cent of the respondents have knowledge about the objective of the VFDC under the National Afforestation Programme (NAP), which shows that the population is well aware of the need for afforestation. In terms of the functions of VFDC, only 8 per cent of the total respondents had knowledge about the different functions. Regarding the benefits again only 20 per cent of the total population of the respondents were aware of the benefits of afforestation. The response regarding knowledge of Management plan of the afforestation programme was nil. In terms of the Disposal of Forest Produce, the response is again nil, the reason for this is mainly because the plantation were still young and no actual harvest or produce has been benefited from them to be disposed to gain benefit. Also majority of the respondents mainly use forest produce for domestic purpose and not for commercial purpose. An important point noted here is regarding the respondent's knowledge of the role played by Forest Department and the officials, for which the response was only 4 per cent.

Awareness about the Benefits Gained from VFDC

The response options of Yes/No revealed that 88.33 per cent respondents responded that they are aware of the benefits available through VFDC. Only 11.67 per cent of the respondents responded they are not aware of the benefits available through the scheme.

1. The reason for responding Yes
   (i) Through VFDC the members were provided provisions such as small farm tools like plough, spade, machete, tub, gas stove and cylinder, cooker, cooking utensil for YMA etc.
   (ii) Infrastructural developments for the locality were made possible through the EPA funds of VFDC.

(iii) Wage labour opportunity was made available to members through the plantation work in the project.

(iv) VFDC acted as an additional channel for raising fund towards the development of the locality and the people.

2. The reason for responding No

(i) The respondents felt that plantation project was a failure because existing forests were cleared for creation of project site and at times new plants die off due to lack of proper maintenance creating more barren land.

(ii) The wage rate prescribed as per the guideline of the Scheme is too less as compared to the extent of labour involved in the plantation work of the VFDC, so it becomes less preferable for the wage earner to work in such plantation sites.

(iii) The respondents also mentioned that the EPA fund was reduced from the initial ₹ 4000 per ha to only ₹ 1000 per ha which made it less desirable for localities/villages to surrender their land for project sites under VFDC.

Awareness about the Meetings of VFDC

An important function of the VFDC is to conduct regular meetings such as Entry Point Activity, Work meeting and Awareness meetings within the localities which is an important way to communicate to the beneficiaries about the overall working of the VFDC and give them training facilities. Through the fieldwork, it was found that as regards the number of meetings, the members of general body attended, 10 per cent respondents have not attended any meeting, 76.67 per cent have attended 1 to 3 times, 13.33 per cent of the respondents from the general body have attended the meetings 3 to 5 times. The reason for lesser number of meetings held with regards to the general body is because the general body meeting or general meeting is held only at the time of the election of the Managing Committee members which is conducted every two years. When the respondents were asked if the elections were conducted after every two years the respondents mentioned the fact that the initiative for conducting the election of the Managing Committee was in the hands of the Forest Department, therefore, they were not frequently conducted.

Training on Forestry and Plantation Project by the Forest Department

Training facilities form an important component of the functions of the VFDC. In this study it is revealed that 13.33 per cent of the respondents received ground level training but the majority (86.67%) did not receive any training. Based on the response acquired from the schedule of the forest officials, technical training is not provided to the members, only the forest officials are given technical training who then guide and assist the members with necessary technical know-how.

Involvement and Level of Contribution of Members towards VFDC

There are different activities of the VFDC where the beneficiaries can take active role in involving themselves by way of creating nurseries, selecting the forest land, plantation work, and it was found that, only 6.67 per cent were involved in choosing the land or area for creating the plantation project, only 5 per cent of the members were involved in creation of Nursery. Majority of the members/beneficiaries, i.e. 43.33 per cent were involved by means of labour or working in the plantation site, 30 per cent of the members were involved in protection and conservation of the project site, 15 per cent of the respondents although being members did not have any involvement. As regards contribution towards VFDC, majority of the respondents that is 71.67 per cent contribute to VFDC in the form of labour or wage labour in the project site and when labour is required when some activity is done by VFDC within the village for developmental purpose, nil or 0 per cent contribute to VFDC by finance, 15 per cent contribute by any other means in terms of allowing the committee to hold meeting in their home, landing their farmland for project site etc. while 13.33 per cent of the respondents have responded that they have not contributed to VFDC in any way.

Conclusion

A cursory glance at the overall working of the VFDC, with regard to regeneration, management, conservation and protection of forests land, all the three localities reveals that the VFDC has basically functioned as a management strategy focusing on two objectives: (1) Development of forest land, and (2) Village development. All the three localities of the present study have adopted 240 hectare of land for plantation project out of which 60 hectare has been successfully created into plantation lands, although, these plantation projects are at developing stage. The maintenance funds for these plantations are incurred by the government through the FDA which is provided for a period of only three years from the creation of the plantation. After the three years, the initiative for maintaining the plantation is upon the VFDC without the assistance from the FDA. Land as a resource generally belongs to the Village council in Mizoram, therefore, by implementing the schemes of VFDC it has been found that the local people are able to utilize their land resource towards development of their forest land and also their village.

As regards regeneration of these forest land systematic and modern methods for creating plantation have been applied with the active involvement of the local people and the Forest Department, such as pre-set-up models for plantation, creation of soil and moisture conservation in the plantation area, creating fire line for protection against fire, and fencing the plantation for proper protection inducing a sense of ownership to the people towards the forest and a sense of responsibility in afforestation.

Looking at the overall development of the village through the schemes of VFDC, the three villages have gained many personal as well as infrastructural gains. This is evident from the EPA activities which have provided the local community with secondary sources of employment as well as infrastructural development of the village as a whole. This has enhanced a sense of security amongst the local people.

REFERENCES

Anonymous (2001). *Proceedings of the JFM Nodal Officers Meeting*, 5th December, 2001, New Delhi.

Government of India (2007). *State of Forest Report*, New Delhi: Ministry of Environment and Forest.

Government of Mizoram (2008). *Report on the Status of Forest in Mizoram*, Aizawl: Department of Environment and Forest.

——— (2009). *Revised Operational Guidelines of National Afforestation Programme*, Aizawl: Office of the Principal Chief Conservator of Forests.

Brasnett, N.V. (1953). *Planned Management of Forests*, London: George Allen & Unwin Ltd.

Helms, J.A. (ed.) (1998). "Terminology of Forest Science, Technology, Practice and Products", *The Dictionary of Forestry*, Society of American Foresters: Bethesda, Md.

Prakash, Ram (2006). *Forest Management*, Dehra Dun: International Book Distributor.

BIBLIOGRAPHY

Appasamy, Paul P. (1993). "Role of Non-Timber Forest Products in a Subsistence Economy: The Case of a Joint Forestry Project in Indian", *Economic Botany*, 47(3): 258-267.

Arora, Dolly (1994). "From State Regulation to People's Participation: Case of Forest Management in India", *Economic and Political Weekly*, 29(12): 691-698.

Centre for Science and Environment (1985). *The State of India's Environment 1984-85: A Second Citizen's Report*, New Delhi: Centre for Science and Environment.

Darlong, V.T. (2002). "An Overview of Forest Policies and Legislations *vis-à-vis* Forest Resource Management in North-East India", in Datta, Ray, B. Alam, K. (ed.), *Forest Resources in North-East India*, (13-38), New Delhi: Omsons Publications.

Das, Achyut Sri (1997). "Joint Forest Management in Tribal Areas: Conflict and Resolution", in Mohapatra, P.M. (ed.), *Forest Management in Tribal Areas: Forest Policy and Peoples Participation*, New Delhi: Concept Publishing Company.

Datta, Ray and B. Alam, K. (2002). *Forest Resources in North-East India*, New Delhi: Omsons Publications.

Dharia, Mohan (1986). *Afforestation in India*, Pune: Vanarai Trust.

Earl, D.E. (1975). *Forest Energy and Economic Development*, Oxford: Clarendon Press.

Gantayat, S.P. (1997). "A look at Tribal and Forest", in Mohapatra, P.M (ed.), *Forest Management in Tribal Areas: Forest Policy and Peoples Participation*, New Delhi: Concept Publishing Company.

Government of Mizoram (2006). *State of Forest Report 2006*, Aizawl: Department of Environment and Forest.

Government of Mizoram (2008). *Handbook on Village and Population as on 1.2.2008*, Aizawl: Directorate of Rural Development.

Government of India (2010). *Annual Report to the People on Environment and Forests 2009-2010*, New Delhi: Ministry of Environment and Forest.

Guha, Ramachandra (1983). "Forestry in British and Post-British India: A Historical Analysis", *The Economic and Political Weekly*, 16(4): 18-25.

——— (ed.) (1994). *Social Ecology*, New Delhi: Oxford University Press.

——— (2000). *Environmentalism: A Global History*, New Delhi: Oxford University Press.

Gupta, S.K. (2002). *Aspects and Sustainability of Joint Forest Management*, Dehradun: M/s. Bishen Singh Mahendra Pal Singh.

Harrison, S.R., Herbohn, J.L., Herbohn, K.F. (2000). *Sustainable Small-scale Forestry Socio-economic Analysis and Policy*, Cheltenham, UK. Northampton, MA, USA: Edward Elgar Publishing Limited, Inc.

Jodha, N.S. (2000). "Joint Management of Forests: Small Gains", *Economic and Political Weekly*, 35(50): 4396-4399.

Kovalli, Shashi (1995). "Joint Forest Management: Superior Property Rights?", *Economic and Political Weekly*, 30(30): 1933-1938.

Kurian, Mathew and Jayesh Bhatia (1997). "Forest Guards as Partners in Joint Forest Management", *Ambio*, 26(8): 553.

Miller, Tyler and G. Andrews, St, Jr. (1988-1989). *Environmental Science: An Introduction*, Belmont, California: Wodsworth Publishing Company.

Nair, Ramachandran, P.K. (1993). *An Introduction to Agroforestry (16-17)*, Dordrecht/ London/Boston: Kluwer Academic Publishers, International Centre for Research in Agroforestry (ICRAF).

Negi, S.S. (1987). *Forest Economics*, Dehradun: M/s. Bishen Singh Mahendra Pal Singh.

——— (1988). *Forest Working Plan*, Dehradun: M/s. Bishen Singh Mahendra Pal Singh.

Rawat, A.S. (1991). *History of Forestry in India*, New Delhi: Indus Publishing Co.

Sen, Sipra (1992). *Tribes of Mizoram: Description, Ethnology and Bibliography (1840-1990)*, New Delhi: Gyan Publishing House.

Shangpliang, Rekha M. (2006). Forest in the Life of the Khasis: A Study in the Role of Forest in Khasi Socio-Economic Structure, unpublished Ph.D. Thesis, Department of Sociology, North-Eastern Hills University, Shillong, Meghalaya.

Singh, Daman (1996). *The Last Frontier People and Forests in Mizoram*, New Delhi: Tata Energy Research Institute.

Singset, Seiboi (2010). *Traditional Forestry Management of the Thadou-Kukis*, New Delhi: Akanksha Publishing House.

Sinha, A.C. (1993). *Beyond the Trees, Tigers and Tribes: Historical Sociology of the Eastern Himalayan Forests*, New Delhi: Har-Anand Publications.

Soni, R.C. (ed.) (1975). *Afforestation in India's Economy*, New Delhi: Director, ESRF, Sapru House.

Stebbing, E.P. (1926). *The Forests of India*, Vols. I, II and III, New Delhi: A.J. Printers Agency.

Upadhyay, Sanjau (2003). "JFM in India: Some Legal Concerns", *Economic and Political Weekly*, 38(35): 3629-3631.

9

Sustainable Use and Conservation of Forest Resources in Meghalaya

Brinda Bazeley

Introduction

Sustainability is a social challenge, a call, a task in progress, a journey to achieve a common set of goals and values. Studies have revealed that various cultures and societies are involved in achieving sustainability. The strategies for achieving sustainable human development at the national level must be formulated to reduce poverty, create employment opportunities and strengthen social integration. The word sustainability gained momentum only after the 1980s. It was declared in the World Summit 2005 that sustainability has 3 basic pillars, i.e. environment, social and economic demands. Of the three pillars, the most important is environmental sustainability. If this is not solved, then is no matter how hard we try, the other pillars cannot be made strong because they are dependent on the greater system they live within the environment. A better sustainable development can lead to healthy eco-system which is very important for survival. Sustainability can also be defined as a socio-ecological process characterized by the pursuit of a common ideal. Sustainability is based on a simple principle: Everything that we need for our survival and well-being depends, either directly or indirectly, on our natural environment. To pursue sustainability is to create and maintain the conditions under which humans and nature can exist in productive harmony to support present and future generations. The United Nations Millennium Declaration identified principles and treaties on sustainable development, including economic development, social development and environmental protection. This in accord with the

United Nations Agenda 21, which specifies **culture** as the fourth domain of sustainable development. **Agenda 21** is a non-binding, voluntarily implemented action plan of the United Nations with regard to Sustainable Development.

The goal of sustainable development as such was formally adopted as a development policy of India in 1992.
In that year the Ministry of Environment and Forests, Government of India submitted to Parliament the National Conservation Strategy and Policy statement of environment and development. Now, one of the key attributes of modern aesthetics is ***preservation of nature.*** Sustaining nature by keeping land under forest cover, conserving land resource and protecting land from soil erosions. As per the National Policy 1998, a minimum of $^1/_3$ of the land area of the country should be under forest or tree cover. In hilly areas $^2/_3$ of the land should be under forest cover. Natural resources are derived from the environment. Many of them are essential for our survival while others are used for satisfying our wants. But in recent years, our Natural Resources were slowly vanishing by means of people who abuse it because they do not know the limitations of it. Abusing it can cause food shortage and can also cause Global Warming. Well-being of the society needs sustenance especially in restoring climate change and avoiding natural calamities. Sustainable living is a very important idea and should be the basis of our living standards. We need to stop all kind of wastage. We need to understand that our Natural Resources are limited and overexploitation will harm not only us but the coming generation more. There are simple ways to help in conserving our natural resources. These are the "3R Concept": 3R means "*Reduce*", "*Reuse*" and "*Recycle*". Natural Resources are essential for our survival and are used for satisfying our wants. But our natural resources are reducing because of us, by abusing it and wasting it.

Sustainability is also a social challenge, a call, a task in progress, a journey to achieve a common set of goals and values. Studies have revealed that various cultures and societies are involved in achieving sustainability.

William Wordsworth in his poem:

An Ode to Nature beautifully describes the environment as a learning resource and quoted "*One impulse from the Vernal Wood may teach you more of man of moral, evil and of good than all the sages can".*

The State of Meghalaya is situated in North-Eastern region of India, between the Brahmaputra Valley in the North and Bangladesh in the South. It has a total geographical area of 22,429 sq.kms. It is divided into 7 administrative districts, 30 community blocks and has a total population of 29.64 lakh which is approximately 0.24 per cent of total Indian Population as per the 2011 census. The total estimated forest area of the state is 8,514 square kilometers of which 722.36 square kilometers are directly under the state control. The remaining areas are managed by the district councils of East Khasi hill, Jaintia hills and Garo hills as per the Sixth Schedule of the Constitution of India.

Sustaining Forest Reserves in Meghalaya

The North-East India is divided into three ecological zones with different physiographic climate and socio-economic characteristics. The Himalayan Zone (Arunachal Pradesh, parts of Nagaland, Manipur and Mizoram) has high rainfall and relatively thin population density. The plateau of Meghalaya-Karbi Anglong areas has high rainfall, low hills with moderate population densities. The valleys of Brahmaputra, Barak rivers have high rainfall, low flooded plains, high temperature and humidity with a short dry season and high population density. Meghalaya caters to a large number of evergreen forest and waterfalls and the forest have a unique collection of flora and fauna. Meghalaya is a treasure house of nature with a unique array of vegetations ranging from tropical, sub-tropical and temperate forests. Meghalaya is known for its forest reserves and the traditional practices of preserving them. It is a fact, forest reserves, when they have been created and enforced, have rapidly improved the ecosystem and the economy in general.

Table 9.1: Total Forest Cover in Meghalaya

Sl.No.	*District*	*Area in sq.kms.*	*Forest in Percentage*
1.	East Khasi Hills	2748	35.34
2.	West Khasi Hills	5247	53.52
3.	Jaintia Hills	3819	46.13
4.	West Garo Hills	3714	54.45
5.	South Garo Hills	1850	64.11
6.	East Garo Hills	2603	58.38
7.	Ri-Bhoi	2448	50.24
	Total	**22,429**	

Source: Author's Field Study.

The reserved and protected forests under the control of the State Government are only 993 square kilometers and 8,503 square kilometers is under the district council and private ownership.

There are the three kinds of forests:

(a) state forests,
(b) forests owned by clans, and
(c) private forests.

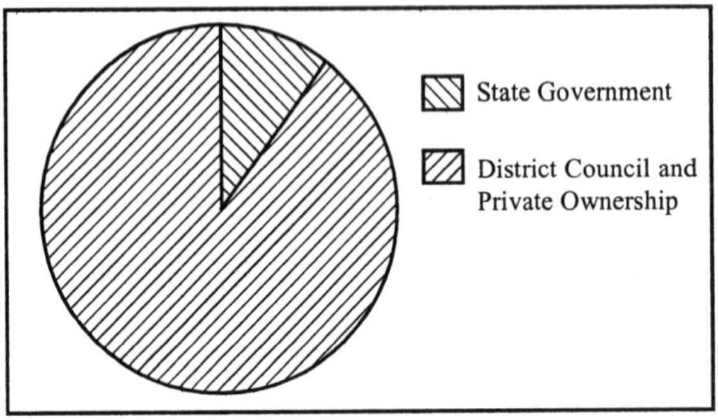

Figure 9.1: Kinds of Forests.

Source: Author's Field Study.

Sustainable Sacred Groves in Meghalaya

Since time immemorial, setting aside patches of forest land and conservation of natural resources has been an integral part of diverse cultures and religious beliefs among different communities. Such forest patches are called "***Sacred Groves***". It has been observed that 1000 sq.km. of forest cover in Meghalaya has remained undisturbed from the time of their existence; these are basically protected by the tribals and are called as "*Sacred Groves*". The sacred groves in Meghalaya largely fall under the temperate watersheds. They are well preserved and play a very important role in sustainability. There are 101 sacred groves in Meghalaya. The most prominent ones are Mawplhang sacred groves found in East Khasi Hills, Lalong Community forest in Jaintia Hills. Garo Hills has 8 sacred groves. Traditional Garos also worship a number of gods named as Tatara Rabugu, A songtatta, Jaripaka etc. for protecting their forests. The sacred groves carries with them strong legends, myths and an indefinable link for centuries, which as a result had a major impact on *sustainability*.

Sacred groves house some very rare endangered, threatened and endemic forest species. The sacred groves are managed by the respective councils, clans and are highly sustainable. The sacred groves have evolved through millions of years. It is reported that only 1.3 per cent of the total sacred grove area is undisturbed and 42.1 per cent area relatively has dense forest, 26.3 per cent had sparse canopy cover and 30.3 per cent has open forest.

It has also been reported that 4 per cent of the total plant species found in Meghalaya are confined to sacred groves.

Management of Sacred Forests in Meghalaya

The Sixth Schedule of the Constitution of India confers power to the tribals of Meghalaya to govern themselves through autonomous district councils in order to protect the tribal rights and ownership including their culture. The rules in respect of management of sacred forest ***Law Lyngdoh, Law Kyntang*** and ***Law Niam*** state that no timber of these forest produce shall be removed for sale, trade or business, but if any timber or forest produce is required for religious purposes a free permit for the purpose shall be obtained. This is done with the recommendation of Lyngdoh with other respective heads. Similarly no trees in protected village/community forest namely Law Sdong and Law Shnong shall be felled or removed. It is a hideous crime and the defaulter can be punished or chased away from the community and locality. The raid land can protect any forest or green block. Therefore, sacred forests have contributed a lot towards sustaining the environment and have close linkage with local ecological values for example medicinal plants of traditional importance which can heal cancer and other dangerous diseases. Besides, the scared groves provide a number of eco-system services such as reduction of soil erosion, maintenance of hydrological cycle, availability of water, dispersal of seeds, reduce habitat destruction, conserve the pollinators and predators, colonization of waste land, conserve the indigenous flora and fauna. These groves are relics of original forests and are store houses of variety of plant genetic resources.

Some Important Sacred Grove sites in Meghalaya are:

Raliang scared grove	Shangpung
Khlaw Blei	Jowai
Phampdem	Nonglyngdoh
Cherrapunji	Mawsmai
Mawphlang	Mawnai
Mawlong	Nonglynkien
Nonglang	Mariam

Nongbri	Iawen
Mawlot	Wanniang Sawkpoh
Diengliengbah	Wahlang Nongkhlung

The state is blessed with best suitable climatic conditions for bio-diversity which supports a large number of flora and fauna species. There are also a large number of rivers which add to the survival of the flora and fauna. The floral diversity includes a good number of trees and shrubs, parasites, medicinal and succulent plants, 325 varieties of orchids only found in Meghalaya. The faunal diversity consists of various types of reptiles, mammals, insects and birds. Meghalaya has around 250 species of colourful butterflies. The bio-diversity promotes a perfect ecological balance.

Nokrek Biosphere Reserve is a national park located approximately 12 kilometers from Tura peak in West Garo Hills district of Meghalaya. The Nokrek National Park has been established at Nokrek and it abounds in various wildlife including herds of wild elephants, rare varieties of birds and pheasants, besides rare orchids. It is a home of Red Panda which has generated curiosity across the world. The park is also home to a very rare species of citrus-indica endemic to this place which the locals call *memang narang* (orange of the spirits). Nokrek is also believed to be the home of *Mande Burung* (jungle man or ape man) and reported cases of sightings abound in and around the villages of Nokrek. It is a valuable ecosystem and is a typical landscape which preserves diverse species. The Nokrek biosphere reserve extends to around 4,748 square kilometers in area. It is the first biosphere reserve of its kind in the North-Eastern region. The UNESCO included this national park in its list of biosphere reserves in May 2009.

This Biosphere Reserve has 3 Basic Functions:

1. *Conservation and Sustainability* — to conserve and sustain the ecosystem, landscape, various species and genetic variation of habitation.
2. *Development* — to foster economic and human development.
3. *Logistics* — to provide logistic support of research, education, both at the national and international level and also to exchange global issues related to conservation and development.

This biosphere reserve can be used to guide and reinforce projects so as to ensure environmental sustainability. It can also attract additional funding from UNESCO and other NGOs, so as to demonstrate approaches to conservation and sustainable development. Nokrek biosphere reserve has its own system of governance to ensure its needs, functions, protection and objectives. Though UNESCO has a major role to play in sustaining and funding. Nokrek biosphere reserve in Meghalaya if successfully satisfies the world heritage criteria in sustaining and protecting the environment it can be complemented to be a world heritage site.

Selbagre Hoolock Gibbon Reserve is located in West Garo Hills District. This small area protected by the community is home to the Hoolock Gibbons, the only ape species found in India. The sacred grove in the reserve is the pride of the village and the community. The Garos never kill or hunt the Hoolock Gibbon as it is a traditional belief among them that if a Hoolock Gibbon is killed, a famine or a curse would befall the entire village.

Balpakram National Park is located in South Garo Hills District. It is a national wildlife park, about 167 kms. from Tura. It is home to one of the rarest animals in the world – the Lesser Panda or the Red Panda as it is commonly known. Balpakram (a place of perpetual winds) covering an area of about 220 sq.km. with a variety of wildlife including tigers, elephants, bison, black bear, leopards, sambar deer, is situated near Baghmara in South Garo Hills district of Meghalaya. The western part of the park adjoins with Siju Bird Sanctuary along the banks of Simsang river.

Baghmara Reserve Forest is located in South Garo Hills District. At a distance of about 4 kilometers from the main town of Baghmara, this thick forest is inhabited by *langurs*, birds of different species, elephants etc. It is a tourist spot and many tourists have reported spotting herds of elephants as they travel from Baghmara to Balpakram National Park via Baghmara Reserve Forest.

Siju Bird Sanctuary is located in South Garo Hills District. Close by on the other side of Simsang River is the Siju Bird Sanctuary, a home for many rare and protected birds and other wildlife. The Siberian ducks migrate here during the winter months. Grey Hornbills are also seen around Siju. One interesting and rare bird is the Peacock Pheasant seen in Siju. November to March is the ideal time to visit Siju.

Nongkhyllem Wildlife Sanctuary is located in Ri Bhoi District. The Nongkhyllem Sanctuary near Nongpoh is also a good place for bird watching. The best time for bird watching is March/April during the nesting season.

Sustainability through Traditional Agricultural Practices in Meghalaya

Meghalaya experiences two major seasons – the winter and the monsoon. It is seen that agriculture is the main occupation of the people of Meghalaya; about 83 per cent of the state depends on agriculture for livelihood. However, only 48 per cent of the land is accounted for agriculture. The ethnic or the tribal communities of Meghalaya follow two major types of Agricultural practices – *jhum* or shifting cultivaticn and terrace cultivation.

Shifting Cultivation in Meghalaya — Also known as slash and burn agriculture it is practiced widely across the state and is the most prevalent form of agricultural practices among the ethnic people and is commonly known as the *Jhum* cultivation; which is the most ancient form of farming believed to have originated.

A piece of land is selected, where the trees and bushes are cut down and burned, in this cleared land seeds or crops are planted without much effort, when the yields slow down, the cultivator moves on to another patch of land and allows the former land to rejuvenate back again naturally, after few years the same land is used again for cultivation.

This tradition has been going on for ages and is the most primitive form of cultivation of agricultural crops and it is time tested. This method has an in-built system of *sustainability and conservation of the environment in the state.*

Terrace or Bun Cultivation — This cultivation dominates on the foothills, slopes and valleys. This system is basically used to provide improved production, conserve soil moisture and prevent land degradation and soil erosion. Here, basically bench or step terraces are constructed on hill slopes, by the cut and bun method. Such measures help to prevent soil erosion and retain maximum rainwater. These traditional cultivations are well adapted to environmental conditions and are highly sustainable. The traditional knowledge among various indigenous communities has enabled them to maintain an ecological balance.

Other Sustainable Practices — *Bamboo based irrigation practices:* It is one of the most widely accepted forms of man's intelligence and skill and has been in practice since time immemorial specially in the Jaintia Hills of Meghalaya. Here, bamboos do the same work as water pipes. This bamboo drip irrigation system is in practice using locally available bamboo species.

The indigenous farmers of Jaintia Hills basically form a network of bamboos and water is carried through these bamboo culms in the indigenous farming areas. This ultimately helps in conserving the environment by saving water and other natural resources to move away from the hilly terrain of Meghalaya.

Traditional Tree Based Farming Practices

This is also a long tradition among the indigenous people of Meghalaya. Tree species are raised from medicine, food, fiber and other agricultural produce. This traditional tree based farming helps in conserving and sustaining the environment. There are various examples of the tree based agricultural farming for example –

Bamboo Farming — This is extensively used for construction, manufacturing, irrigation and young bamboo shoots are also edible.

Multipurpose Trees and Shrubs — Multipurpose farming is a common practice among the indigenous communities of Meghalaya; and are mostly sustained for medicinal purpose, commercial purposes etc.

Some trees like arecanut, bamboo and banana are also planted in the tea gardens so as to provide shade to the tea shrubs.

The homestead farming practice is done at the foothills of the Jaintia Hills and Garo Hills of Meghalaya, small plants like black pepper, ginger, turmeric are successfully grown. The Khasi pine trees are also used to provide moisture to the soil. Various species of ginger, turmeric; vegetables are grown in its shade. These common tradition-based farming practices are found among the indigenous communities of Meghalaya who are not acquainted with any kind of modern technical facilities.

Sustainable Harvesting and Storage

The indigenous communities also have a unique and traditional method of harvesting. The traditional storage system is provided with well ventilated raised bamboo structures; covered with thatched grass. This storage structures are sustainable for high humid conditions. Farmers also use some indigenous insect repellents, *neem* leaves, tobacco leaves for preservation of grains and vegetables. Most of the storage devices are made from bamboo and wood; at times they are plastered with cow dung and mud. Unhusked paddy is always preserved on a elevated level, a little lower than the ceiling of the house constructed by bamboo poles so as to protect from damage by animals especially rats, a covering thatched grass is placed on top to protect from cobs, dust etc. Most of the farmers store the grains in unhusked conditions so that it can be stored for longer durations.

Suggestions for Conservation

1. Possible application of biotechnology in multiplying the threatened species may be explored.
2. Management of the community forests may be improved through adequate funding from the government/NGOs.
3. Separate guidelines should be initiated for the conservation of sacred groves so that the village level institutions are kept at center stage.
4. Continuing research, information management and use of appropriate technologies for updating the sacred groves as traditional preservers lack modern expertise and scientific understanding.
5. Traditional methods coupled with modern technology like ICT can bring sustainability, enhance production and livelihood.
6. In Meghalaya, major forests are with the community and resources with the government. Conservators in the government work as extension workers only and do not show responsibility towards the community forests.
7. Training programmes in management is required for capacity building.
8. Researches in regeneration ecology should be undertaken with respect to endangered species.
9. Raise awareness amongst the villagers regarding modern and scientific conservational techniques.
10. A "scared grove conservational fund" should be instituted to finance the management and protection of the Groves.
11. The fund may be provided by the Global Environment Facility (GEF) or by the Government of India under the protected area network programme.
12. The villages successfully protecting the groves should be given some additional funds for community development.

Conclusion

In sum, the principle of sustainable development has become a guiding principle for management of renewable resources such as forest reserve in Meghalaya. The Brundtland Commission while defining sustainable use of resource emphasized the futuristic demands and needs. "The development that meets the needs to the present without compromising the ability of the future generations to meet their own needs", is called sustainability. The chapter has made an attempt to present a number of aspects of sustainability in Meghalaya forest covers, biosphere reserves and agricultural practices. Thus, portraying the preservation and sustainability of natural environment with all seriousness and at any cost.

REFERENCES

Cowie Jonathan (2007). *Climate Change – Biological and Human Aspects*, Cambridge University Press, UK.

Hardy T. John (2003). *Climate Change Causes, Effects and Solutions*, John Wiley & Sons Ltd., West Sussex PO19 8SQ, England.

Rogers John J.W. and Fiess Geofferey P. (1988). *People and the Earth – Basic Issues in the Sustainability of Resources and Environment*, Cambridge University Press, UK.

Sarkar, A.N. (2010). *Global Climate Change – Beyond Copenhagen*, Pentagon Press, New Delhi.

http://www.thwink.org/sustain/glossary/EnvironmentalSustainability.html

https://en.wikipedia.org/wiki/Sustainability"cite_note-Dr._Joel_Diemer-3- wandemberg, JC (August 2015). *Sustainable by Design.* Amazon. p. 122. ISBN 1516901789. Retrieved 16 February 2016.

*https://en.wikipedia.org/wiki/*Sustainability – United Cites and Local Governments, *Culture: Fourth Pillar of Sustainable Development.*

https://en.wikipedia.org/wiki/Agenda_21 "What is Agenda 21?". ICLEIUSA. Retrieved 8 December, 2012.

http://www-chrisanne.blogspot.in/2010/12/why-do-we-need-to-conserve-our-natural.html

https://en.wikipedia.org/wiki/Meghalaya

http://megtourism.gov.in/wildlife.html

https://en.wikipedia.org/wiki/Baghmara, _India

10

Environmental Worldview among the Teacher-Trainees of DIETs, Meghalaya

Yodida Bhutia and *Georgia Liarakou*

Introduction

A worldview is a mental model of reality, it is a framework of ideas and attitudes about the world, ourselves, and life, a comprehensive system of beliefs. It is a hypothesis we have formed over the years through our experiences, interaction with others and things through various medias, peers, education at home, school and colleges. The worldview can be of any area such as environmental worldview. So, this view is the basic idea through which individual understands the natural world and see relationship between different beings of the natural world. It consists of three components – how people think the world works, what they think their role in the world should be, and what they believe is right and wrong environmental behaviour (Miller, 2003). Environmental worldviews, or environmental beliefs, refer to an underlying belief or value system through which an individual understands the natural world and the ways in which humans are related to, or separate from it (Liu & Lin, 2014). The environmental worldview can also be classified based on the locus of control such as anthropocentric (human-centred), biocentric (life-centred), or ecocentric (earth-centred) (Miller, 2003; Merchant C., 1992; Plumwood, 2007). Anthropocentric worldview is based on the view that humans are the central or most important element of existence as opposed to God or nature. This view considers human as masters of nature. They view that humans have intrinsic value because they exist. However, nature has instrumental value, which is determined only by its usefulness to humans. The biocentric worldview emphasize that all organisms on earth are valuable and all species have intrinsic value and all forms of life have an inherent right to exist. They consider that the protection of all species is more important for the life to continue. The ecocentric worldview demands that nature as a whole is valuable above any of its individual parts. It emphasizes the importance of protecting the ecosystems in which many species live, by protecting ecosystem and premature extinction of species is prevented. These environmental worldview develop with exposure to environmental knowledge through cultural practices and also formal education in schools and colleges.

Education is seen as the great source of knowledge in Indian culture. Today the tools of citizenship or rather the ability to understand and master the important questions of the modern world, are primarily provided by education (Filho, 2006). Therefore, if we want to bring any change in society in relation to knowledge, attitude, skills and behaviour towards environment, it should be environmental education provided through all formal and non-formal medium. Education can play an important role in motivating and encouraging towards protection and conservation of environment. Tbilisi Report in 1977 stated that "the role of education in the face of environmental problems and opportunities is, therefore, a crucial one" (UNESCO, 1977). Report of the United Nations Conference on the Human Environment in 1972 is one of the principle for the preservation and improvement of human environment, states that "education in environmental matters, for the younger generation as well as adults, ...is essential in order to broaden the basis for an enlightened opinion and responsible conduct by individuals, enterprises and communities in protecting and improving the environment in its full human dimension" (United Nations, 1972). Education has impact on the environmental worldview, attitudes and belief (Ruff and Oslon, 2009). And for disseminating this knowledge, teachers do play an important role, for which education provided to teacher educator should be of quality as these prospective teachers have responsibility to teach younger generation. Teacher training programme throughout the world has been given importance as it prepares the teachers for future generation. Teachers need to acquire a solid grounding in the knowledge and skills necessary to effectively provide education. In this pluralistic society where media has penetrated so much, the teacher has to have firm foundation of knowledge of philosophy, psychology and sociology. They play a pivotal role in education and training the young minds. United Nations in 1972 World Conference recommended that training and retraining of teachers on environmental issues are important

(UN, 1972, recommendations 96). It is essential that teachers possess the ability to teach sound environmental values to students from diverse backgrounds as stated in guidelines given by UNESCO in 1985. UNESCO in 1985 in a guide on environmental education stated that environmental educators are chief among other agents who seek to accomplish the difficult task of strengthening, reinforcing or human-made environments, it is virtually impossible to avoid confronting the personal values of learners that form the basis of their actions. This fact has caused environmental values education to be of major importance in the general environmental education curriculum. Teachers play a key role in environmental literacy of future generations (UNESCO, 1987; McKeown & Hopkins, 2002). In 1987 Brundtland Report, also reported that a critical point of intervention is during teacher training. The attitudes of teachers will be key in increasing understanding of the environment and its links with development. Their environmental worldview is very important. The future quality and stability of life on our planet depends on youngsters developing the worldview necessary for making informed and sensitive decisions about the environment and becoming active participant in the creation of sustainable world (Erdogan, 2013). Therefore, this chapter is focused on prospective teachers who are undergoing the training course in the District Institute of Education and Training (DIET), Meghalaya, India.

When we consider man, then it is imperative we should not separate man and culture. Popularly culture is said to be the characteristic and knowledge of a particular group of people, which includes language, religion, cuisine, social habits, music and arts. Cultural beliefs play a large role in determining environmental attitudes as stated by Alvin (1976) as quoted in (UNESCO, 1985). The human attitudes and belief which forms the worldview are certainly impacted by the culture in which we grow and live. In this connection, the human being cannot have worldview solely based on one aspect rather their view will be affected by their culture with which they have grown. To comprehend the deep meaning of Environment, the anthropogenic environment ought to be included in the definition (Tsekos, 2009). Therefore, this research focused on gender and religion among teacher-trainees and find out its impact in their worldview.

Research in Teacher Education Programme

The study is based in the District Institutes of Education and Training (DIETs), Meghalaya, India, which are involved in training the prospective teachers for the elementary schools and primary schools and candidates required to have completed high school to be eligible for the course. In India, the DIETs are located in each district of the State, which provides training to teachers of primary and elementary schools. Meghalaya has seven DIETs, located in Sohra, Thaladskein, Nongpoh, Nongstoin, Rusubelpara, Tura and Baghmara. Furthermore, in view of the interdisciplinary nature of EE, it can both contribute to and benefit from the integrative nature of pedagogy studies. Finally, the pedagogical component provides a built-in framework for addressing practical aspects of the pedagogies for teaching environmental dilemmas in the classroom. The fact that students (both environment and non-environmental affiliated) toward the end of their studies acknowledge subject-matter as a basis for relevance, supports that teacher-preparation courses should be encouraged to integrate EE into all teaching subjects (Yavetz, Goldman, & Peter, 2014). Xiao & Hong (2012) found that higher level of knowledge associated with greater concern for the environment among urban Chinese students.

Purpose

The purpose is to study the Environmental Worldview of teacher-trainees of DIETs, Meghalaya. The purpose of the study are put in the form of following research questions:

1. What is the environmental worldview of teacher-trainees of Meghalaya, India?
   (a) Does gender relate to teacher-trainees' environmental worldview?
   (b) Has religion made impact on the teacher-trainees' environmental worldview?
2. Do the teacher-trainees hold traditional belief on protection of nature?
3. What is the source of environmental values among the teacher-trainees?
4. How do they view themselves as the teachers to teach environmental values?

Method

Population and Sample

The population of the study includes the teacher-trainees of the DIETs,

Meghalaya. The total population in all seven DIETs of Meghalaya, India were 398. The sample of the study includes the 198 teacher-trainees of DIETs, Meghalaya and random sampling technique was used to collect sample from the population.

Research Design

The study adopted survey method as have been adopted by Ewert and Baker in 2001, Nooney, Woodrum, Hobas, & Cifford in 2003, Ruff and Oslon in 2009, Taskin in 2009, Liu and Lin in 2013, Noblet, Lindenfeld, & Anderson, 2013, Erdogan, 2013. It is descriptive and explanatory in nature. This method suits this as the teacher-trainees of DIETs, Meghalaya as they can read and can rate their opinion themselves. Survey research is considered method of systematic data collection, though, the survey has a long historical tradition. As far back as the time of the ancient Egyptians, population counts and surveys of taxes etc. (Borg & Gall, 1989).

Instrumentation

The study used New Ecological Paradigm Scale developed by Dunlap *et al.*, 2000. According to Dunlap (2008), this scale measures "the degree to which respondents view the world ecologically". Lower scores are indicative of a more ecological worldview, whereas higher scores are indicative of a more anthropocentric worldview. The NEP scale consists of 15 five-point Likert-type items, ranging from Strongly Disagree to Strongly Agree. Seven items are reverse-scored (indicating anthropocentrism). The scores can range from 15 to 75. Noblet, Lindenfeld, & Anderson, 2013 cited that odd numbered items are reverse coded, the scores for all items are summed, and then a mean score is calculated for each individual.

Results

The Environmental Worldview of Teacher-Trainees of DIETs, Meghalaya

The data on the environmental worldview has been calculated using frequency as well as mean and standard deviation.

Majority 50.51 per cent Indian teacher-trainees have scores in environmental worldview is between 39–50, 43.94 per cent Indian teacher-trainees have scores in environmental worldview is from 27–38, only 5 per cent have 51–62 and 0.5 per cent Indian teacher-trainees have scores 15–26. This indicates the teacher-trainees have more ecological worldview.

Table 10.1: Frequency and Percentage of Teacher-Trainees Scoring Different Level of Environmental Worldview

Scale for Environmental Worldview	*Teacher-Trainees*		*Level*
	Frequency	*Per cent*	
63–75	—	—	High
51–62	10	5.05	Above average
39–50	100	50.51	Average
27–38	87	43.94	Below average
15–26	1	0.5	Low

Note: Lower the score-ecological worldview, higher the score-anthropocentric.
Source: Field Study.

Gender Difference in Environmental Worldview

In order to find out the difference between male and female teacher-trainees of India in their environmental worldview a null hypothesis was formulated, H_0: There is no significant difference in the environmental worldview between male and female of teacher-trainees of Meghalaya, India.

Table 10.2: Mean, Standard Deviation (SD) and *t*-Value of Male and Female Teacher-Trainees on their Environmental Worldview

	Gender	*N*	*Mean*	*SD*	*t-Value*
DIETs teacher-trainees	Male	37	40.73	6.91	1.12*
	Female	161	39.45	6.15	

$P > 0.05$.
Source: Field Study.

The mean score of Indian female and male teacher-trainees in environmental worldview are 39.45 (6.91) and 40.75 (6.15). Female teacher-trainees have lesser mean score than their male counterparts showing that female have inclinations towards ecology and its protection. Independent sample *t* test was conducted to compare environmental worldview of male and female teacher-trainees of DIET. There was no significant difference between male (M = 39.45, SD = 6.91) and female (M = 40.75, SD = 6.15), $t\,(196) = 1.12, p = 0.265$. These result suggest that gender does not effect environmental worldview among teacher-trainees of DIETs.

Environmental Worldview in Relation to Religion

The study considered religion is one of the important variable which may affect the collective on environment so the level of religiousness among teacher-trainees were found and their environmental worldview for which mean and standard deviation was calculated.

Table 10.3: Mean, Standard Deviation on Environmental Worldview of Teacher-Trainees with different Level of Religiousness

Level of Religiousness	*Indian Teacher-Trainees*		
	N	*Mean*	*SD*
High religious	3	39.33	5.69
Above Average religious	7	37.43	4.72
Average religious	68	39.82	5.99
Below Average religious	107	39.79	6.60
Low religious	13	39.38	6.84

Source: Field Study.

It is seen from Table 10.3, that Indian teacher-trainees, majority consider themselves below average religious (54%), average (34.3%), low religious (6.6%), above average (3.5%) and high (1.5%). Mean score of teacher-trainees with all level of religiousness from low to high has mean score ranging from 37.43-39.82, which shows that despite having different level of religiousness among teacher-trainees, the mean score in environmental worldview is not differing much and has remained ecological worldview.

Table 10.4: Pearson Product Moment Correlation between Religion and Environmental Worldview

		DIET Teacher-Trainees
Religious	Pearson Correlation	1
	Sig. (2-tailed)	
	N	198
Environmental worldview	Pearson Correlation	0.02
	Sig. (2-tailed)	0.74
	N	198

Source: Field Study.

A Pearson product moment correlation was computed to assess the relationship between the religion and environmental worldview among teacher-trainees of there is no relationship between the two variables $r = 0.02$, $p = 0.74$, $n = 198$. Whether the teacher-trainees are highly religious or least religious, do not have effect on the environmental worldview among teacher-trainees of DIET.

Family as Source of Environmental Worldview

Table 10.5: Frequency, Percentage of DIETs Teacher-Trainees on Family as Source of Information on Environment

	DIET Teacher-Trainees	
	Frequency	*Per cent*
High	25	12.6
Average	151	76.3
Low	22	11.1
Family is not the source	—	—

Source: Field Study.

Table 10.5 shows that among DIETs teacher-trainees majority (76.3%) mentioned that home is the average source of information on environment. And 12.6 and 11.1 per cent mentioned that family was high and low source of information on environment and its protection.

School as Source of Environmental Worldview

Table 10.6: Frequency, Percentage of DIETs Teacher-Trainees on School as Source of Information on Environment

	Indian Teacher-Trainees	
	Frequency	*Per cent*
High	122	61.6
Average	72	36.4
Low	4	2.0
Not source	—	—

Source: Field Study.

When asked about School as the source of information, majority (61.6%) of Indian teacher-trainees mentioned that school was the high source of information on environment, 36.4 per cent consider school as average source of information and negligible (2%) mentioned that school was low source of information. Here also school was major source of information for Indian teacher-trainees, on environment and its related issues.

College/University as Source of Environmental Worldview

DIETs teacher-trainees mentioned that college is the average source of information on environment by majority (53.5%) and as high source by 43.9 per cent and negligible (2.5%) mentioned that college was the low source of information.

Table 10.7: Frequency, Percentage of DIET Teacher-Trainees on College/University as a Source of Information on Environment

	Indian Teacher-Trainees	
	Frequency	*Per cent*
High	87	43.9
Average	106	53.5
Low	5	2.5
Not source	—	—

Source: Field Study.

Internet as Source of Environmental Worldview

Table 10.8: Frequency, Percentage of DIET Teacher-Trainees on Internet as a Source of Information on Environment

	Indian Teacher-Trainees	
	Frequency	*Per cent*
High	79	39.9
Average	74	37.4
Low	44	22.2
Not source	1	0.5

Source: Field Study.

The study found out that among DIET students, 39.9 per cent mentioned that internet is the higher source of information on environment, 37.4 per cent consider it as average and 22.2 per cent as low source of information on environment.

Environmental Educator

Teacher-trainees of DIETs were asked how do they rate themselves and four options were provided. The result has been calculated using frequency and percentage.

Among the teacher-trainees, 7.6 per cent consider themselves as a very good teacher and 43.9 per cent consider they are good teacher of environmental education, they think they are ready to impast knowledge, and inculcate attitude, skills and values related to environmental education. However, 41.4 per cent mentioned that they can improve with further input. Still 6.6 per cent think that they are unprepared to teach the subject of environmental education.

Table 10.9: Frequency and Percentage of Teacher-Trainees on How do They View as Teacher of Environmental Education

	DIET Teacher-Trainees	
	Frequency	*Per cent*
Very good	15	7.6
Good	87	43.9
Improved with further input	82	41.4
Unprepared	13	6.6
Did not reply	1	0.5

Source: Field Study.

Among those teacher-trainees who want to improve with further input mentioned that 47.6 per cent would like to attend some workshop on environment and 52.4 per cent would like to attend separate course on environmental education.

Discussion and Conclusion

1. The study found out that both DIETs teacher-trainees have more ecological environmental worldview. These teacher-trainees share a collective view that nature is important part of the life and they feel the need to protect the environment in which many living organisms live and prevent extinction of many species. They do not agree with view that the human beings are masters of nature and they can use natural resources for the benefit of human being. Similarly, in another comparative study by Petegem & Blieck, 2006 found that respondents in Belgium believe in human-nature equality, whereas Zimbabwean respondents feel more dominant over nature and emphasize a utilitarian view of the environment.
2. The study found that gender do not impact the environmental worldview of teacher-trainees of DIET. Similar findings were made (by Smith, 2011) that there were no differences in NEP scores by gender (Noblet, Lindenfeld, & Anderson, 2013). Women faculty members reported statistically higher NEP scores than men faculty members, however, gender differences do not exist in the population sample in environmental worldview. However Nooney, Woodrum, Hobas & Cifford

(2003) found that women scored higher in worldview scale of NEP (Tsachalidis, Tampakis, Tsantopoulos, Zografou & Arabatzi, 2006). There was significant difference found to exist between gender categories and teachers' knowledge. In particular men have better knowledge than women.

3. There is no significant difference between different level of religiousness and environmental worldview. Similarly Nooney, Woodrum, Hobas & Cifford (2003) also found that there was no differences found between those with no affiliation and religious moderates or fundamentalists in their environmental worldview. The present study also found that there is no relationship between religion and environmental worldview.
4. More of DIET teacher-trainees confirmed that family, school, college have been a source of information on environment and its related issues as opposed to Greek teacher-trainees. This indicate that for Indian teacher-trainees idea of environment is influenced and has received knowledge, skills, and attitudes by observing elders in the family, from the teaching of schools and colleges. The study also found that internet has not been major source of information related to environment, its issues and its protection and conservation. Despite more internet are available but it has not categorically made impact on the view of environment of the young prospective teachers of both the countries.
5. The study found that more of DIET teacher-trainees do not consider themselves as good and ready to teach environmental education in schools. Among them 52.4 per cent mentioned that they would like to attend separate course on environmental education and 47.6 per cent said that they would like to attend some workshop on environmental education.

Implications

Our results indicate despite the teacher-trainees have ecological worldview have expressed the desire to learn more about environment through workshops and even a course on environmental education. Our results also suggest that majority of DIET teacher-trainees have stated the need of a course on environmental education so this finding will help District Educational Research and Training (DERT) which looks after the training of teachers in all DIETs of Meghalaya, India to formulate a course or training programme on environmental education. They also feel less confident as environmental education teacher so there is a need of more training on knowledge, skills, values and attitude towards environment so that they become better teacher to manage the subject on environmental education in schools.

Acknowledgements

The authors would like to gratefully acknowledge funding support by State Scholarship Foundation (IKY), Greece for granting Post-doctoral scholarship to Yodida Bhutia, North-Eastern Hill University (NEHU), Shillong, Meghalaya, India. She worked with Prof. G. Liarakou, University of the Aegean, Rhodes, Greece. We also thank all the teacher-trainees who participated in this research.

REFERENCES

Basu, R.L. (n.d.). *Environment and Ecology in Kautilya's Arthashastra.* Retrieved April 27, 2015, from *www.academia.edu: http://www.academia.edu/7781197/Environment_and_Ecology_in_Kautilya_s_Arthashastra*

Black, J. (1970). "The Dominion of Man", in R.J. Berry, *Environmental Stewardship* (pp. 100-150). Edinburgh: Bloomsberry Academic. Retrieved May 13, 2015, from: *https://books.google.gr/books?id=-r4c4hSaA6UC*

Bodnar, I. (2012). *Aristotle's Natural Philosophy.* Retrieved April 26, 2015, from: *plato.stanford.edu/archives/spr2012/entries/aristotle-natphil: http://plato.stanford.edu/archives/spr2012/entries/aristotle-natphil*

Borg, W.R. and Gall, M.D. (1989). *Educational Research*, New York: Longman Inc.

Branton, C.J. (2006). "Indigenous Traditions and Ecology", *Environmental Practice*, 8(4).

Datta, D. (1936). "The Moral Conception of Nature in Indian Philosophy", *International Journal of Ethics*, 46(2): 225-226. Retrieved from: *www.jastor.org.*

Dunlap, R. (2008). "The New Environmental Paradigm Scale: From Marginality to Worldwise Use", *Fall*, 40(1): 1-18.

Dunlap, R., Liere, K.D., A.G. Mertig, and Jones, R. (2000). "Measuring Endorsement of the New Ecological Paradigm: A Revised NEP Scale", *Journal of Social Issues*, 56(3): 425-442.

E.M. (1999). *Seven Complex Lessons in Education for the Future*, Paris: UNESCO.

Erdogan, N. (2013). "Environmental Worldview in Higher Education: A Case Study of Turkish College Students", *Procedia – Social and Behavioural Sciences*, 106: 1086-1095. *doi:10.1016/j.sbspro.2013.12.122*

Ewert, A., and Baker, D. (2001). "Standing Where You Sit: An Exploratory Analysis of the Relationship between Academic Major and Environmental Beliefs", *Environment and Behaviour*, 687-707.

Filho, W. (2006). *Innovative Approaches to Education for Sustainable Development*, Frankfurt: Peter Lang GmbH.

Giannasi, M. (2012). "Nature is the Realm of the Incomprehensible", in M.F. Plebani, *Existence and Nature: New Perspective* (pp. 100-102), Berlin: De Gruyter.

Gosling, D.L. (2001). *Religion and Ecology in India and Southeast Asia*, London: Routledge.

Government of India (2001). *Census Data*, Retrieved April 28, 2015, from: *www.censusindia.gov.in/Census_And_You/religion: http://censusindia.gov.in*

Government of Meghalaya (n.d.). *Education in Meghalaya*, Retrieved April 28, 2015, from: *megeducation.gov.in: http://www.megeducation.gov.in*

Hawcroft, L., and Milfont, T. (2010). "The Use (and Abuse) of the New Environmental Paradigm Scale over the Last 30 Years: A Meta-Analysis", *Journal of Environmental Psychology*, 143-158. *doi:10.1016/j.jenvp.2009.10.003*

Hill, P.J. (2000). "Environmental Theology: A Judeo-Christian Defence", *Journal of Markets and Morality*, 3(2): 158-172.

Jucker, R. (2002). *Our Common Illiteracy: Education as if the Earth and People Mattered*, Frankfurt: Peter Lang GmbH.

Kadji-Beltran, C., Constantinou, C., Lindemann-Matthies, P., and Zacharia, Z. (2006). "Diversity in Environmental Education: The Pre-service Training of the Primary School Teachers", in W.L. Filho, and M. Salomone, *Innovative Approaches to Education for Sustainable Development* (pp. 65-71), Frankfurt: Peter Lang GmbH.

Kawagley, A.O. (1999). *Education Indigenous to Place: Western Science Meets Native Reality*, Retrieved May 15, 2015, from: *ANKN website: http://www.ankn.uaf.edu*

Kawagley, A., and Barnhardt, R. (1999). *Education Indigenous to Place: Western Science Meets Native Reality*, Retrieved from: Alaska Native Knowledge Network: *http://www.ankn.uaf.edu*

Kosmopoulos, P., Kantziouva, A., Kosmopoulus, I., and Kleskas, K. & M. (2015). "The Impact of Recession on Public Attitudes toward Energy and the Environment", *International Journal of Sustainable Energy*, 1-18. *doi:10.1080/14786451.2015.1009460*

Krishnamurti, J. (1975, July). *J. Krishnamurti Online*, Retrieved April 24, 2015, from: *www.jkrishnamurti.org: http://www.jkrishnamurti.org*

Lewis, V.B. (n.d.). *Natural Law, Natural Rights, and American Constitutionalism*, Retrieved April 26, 2015, from: *http://www.nlnrac.org/classical/plato: htttp://www.ninrac.org*

Liu, S.C., and Lin, H.S. (2014). "Undergraduate Students' Ideas about Nature and Human-Nature Relationships: An Empirical Analysis of Environmental Worldviews", *Environmental Education Research*, 20(3): 412-429. *doi:10.1080/13504622.2013.816266*

MacDonal, W., and Hara, N. (1994). *Gender Differences in Environmental Concern among College Students*, ASDD Area Sistemi Dipartimentali e Documentali, Università di Bologna, Catalogo delle riviste ed altri periodici.

Martin, M. (2008). *Proclus on Nature*, Retrieved from: Leiden University website: *http://openaccess.leidenuniv.nl*

McKeown, R., and Hopkins, C. (2002). "Weaving Sustainability into Pre-service Teacher Education Programmes", in W.F. Filho, *Teaching Sustainability at Universities: Towards Curriculum Greening* (pp. 251-274), Frankfurt: Peter Lang.

Meghalaya Times (2015, April 28). *Ancient Khasi's Concept of Environment*, Shillong, Meghalaya, India.

Merchant, C. (1983). "Mining the Earth's Womb", in J. Rothschild, *Machina Ex Dea* (pp. 99-116), New York: Pergamon Press.

Merchant, C. (1992). *Radical Ecology: The Search for a Livable World*, New York: Routledge.

Miller, J. (2003). "Daoim and Nature", in H. Selin, *Nature Across Cultures: Views of Nature and the Environment in Non-Western* (pp. 393-409), Dordrecht: Kluwer Academic Publishers.

Naess, A. (1973). "The Shallow and the Deep, Long-Range Ecology Movement: A Summary", *Inquiry*, 16(1): 95-100.

Noblet, C.L., Lindenfeld, L., and Anderson, M. (2013). "Environmental Worldviews: A Point of Contact or Barrier?", *Sustainability*, 5, 4825-4842. *doi:10.3390/5114825*

Nooney, J., Woodrum, E., Hobas, T., and Cifford, W. (2003). "Environmental Worldview and Behaviour", *Environment and Behaviour*, 35(6): 763-783. *doi:10.1177/00139165032562246*

Ormsby, A. (2013). "Analysis of Local Attitudes Toward the Sacred Groves of Meghalaya and Karnataka, India", *Conservation and Society*, 11(2): 187-197. Retrieved from: *http://www.conservationandsociety.org*

Petegem, P., and Blieck, A. (2006). "The Environmental Worldview of Children: A Cross-cultural Perspective", *Environmental Education Research*, 12(2): 625-635. *doi:10.1080/13504620601053662*

Plumwood, V. (2007). "Environmental Ethics", in A.T.J. Pretty, *Environmental Ethics* (pp. 240-266), London: Sage.

R.S. Triparthi (2005). *Sacred Groves of North-East India and Their Floristic Richness and Significance in Biodiversity Conservation*, Lucknow: International Society for Environmental Botanists.

Religion in Greece (2015). Retrieved from: The Greek Island Specialist: *http// :www.greeka.com*

Ruff, C., and Oslon, M. (2009). "The Attitudes of Interior Design Students Towards Sustainability". *International Journal of Technol Des Edu*, 19: 67-77. *doi:10.1007/s10798-007-9038-0*

Sengupta, M., Das, J., and Maji, P. (2010). "Environmental Awareness and Environmental Related Behaviour of Twelfth Grade Students in Kolkata: Effects of Stream and Gender", *Anwesa*, 5: 1-8.

Shiva, V. (1988). *Staying Alive: Women, Ecology and Survival in India*, New Delhi: KALI FOR WOMEN.

Sia Su, G.L. (2008). "Environmental Worldview and Concern of College Students in the Philippines", *Insternational Journal of Sustainability in Higher Education*, 9(1): 39-47.

Smith, E.L. (2011). *The Differences between Students' Knowledge of Environmental Apparel and Environmental Worldview based on College Major and Gender*, Indiana: Unpublished Masters Thesis, University of Ball State. Retrieved from: *http://www.cardinalscholar.bsu.edu*

Smith, H. (2007). *World Wisdom*. Retrieved April 23, 2015, from: *www.worldwisdom.com: http://www.worldwisdom.com*

Stara, K., Tsiakiris, R., and Wong, J.L. (2015). "The Trees of the Sacred Natural Sites of Zagori, NW Greece", *Landscape Research*, 1-23. *doi:10.1080/01426397.2014-911266*

Taskin, O. (2009). "The Environmental Attitudes of Turkish Senior High School Students in the Context of Postmaterialism and the New Environmental Paradigm", *International Journal of Science Education*, 31(4): 481-502. *doi:10.1080/09500690701691689*

Tikka, P., Kuitunen, M., and Tynys, S. (2010). "Effect of Educational Background on Students' Attitudes: Activity Levels and Knowledge Concerning the Environment", *The Journal of Environmental Education*, 31(3): 12-19. *doi:10.1080/00958960009598640*

Tsachalidis, E., Tampakis, S., Tsantopoulos, G., Zografou, D.T., and Arabatzi, Z. (2006). "Teachers and Environmental Education in Greece", *Sustainable Management and Development of Mountainous and Island Areas* (pp. 163-171). Heraklion-Crete: E-media, University of Crete.

Tsekos, C.A. (2009). "Ethics, Science and Environment: The Need for New Environmental Worldview", *International Journal of Environmental Studies*, 66(6): 679-687. *doi:10.1080/00207230903028326*

Tsekos, C.A., and Matthopoulo, D.P. (2009). "Ethics, Science and Environment: The Need for a New Environmental Worldview", *International Journal of Environmental Studies*, 66(6): 679-687. *doi:10.1080/002072309028326*

Tuna, M. (n.d.). *Public Environmental Attitudes in Turkey*, Turkey: Unpublished dissertation: Mugla University.

UNESCO (1961). *Rabindranath Tagore: A Universal Voice*, France: UNESCO.

UNESCO (1977). *Intergovernmental Conference on Environmental Education*, Paris: UNESCO.

UNESCO (1985). *A Guide on Environmental Values Education*, Paris: UNESCO.

UNESCO (1987). *Report of the World Commission on Environment and Development: Our Common Future*, Paris: UNESCO Publishers.

United Nations (1972). *Report of the United Nations Conference on the Human Environment*, Switzerland: United Nations Publication.

Vishwas Rao, A., and Gadgil, A. (2006). "Revitalizing the Philosophy of the *Gurukul* System for Effective Environmental Education", in W.L. Filho, and M. Salomone, *Innovative Approaches to Education for Sustainable Development* (pp. 92-95), Frankfurt: Peter Lang GmbH.

White, L. (1967). "The Historical Roots of Our Ecologic Crisis", *Science*, 155(3767): 1203-1207. Retrieved from: *http://www.jstor.org/stable/1720120*

Wisdom Library (2015, January 9). Retrieved April 23, 2015, from: *www.wisdomlib.org: http://www.wisdomlib.org*

Xiao, C., and Hong, D. (2012). "Gender and Concern for Environmental Issues in Urban China", *Society & Natural Resources: An International Journal*, 25(5): 468-482. *doi:10.1080/08941920.2011.565455*

Xiao, C., and McCright, A.M. (2012). "Explaining Gender Differences in Concern about Environmental Problems in the United States", *Society & Natural Resources: An International Journal*, 25(11): 1067-1084. *doi:10.1080/08941920.2011.651191*

Yao, X. (2000). *An Introduction to Confucianism*, London: Cambridge Press.

Yavetz, B., Goldman, D., and Pe'er, S. (2014). "How Do Preservice Teachers Perceive 'Environment' and Its Relevance to their Area of Teaching, *Environmental Education Research*, 20(3): 354-371. *doi:10.1080/13504622.2013.803038*

Yimaz, O., Boone, W., and Andersen, H. (2004). "Views of Elementary and Middle School Turkish Students Towards Environmental Issues", *International Journal of Science Education*, 26(12): 1527-1546. *doi:10.1080/0950069042000177780*

11

The Impact of Border Haats on the Socio-Economic Life of the People of Meghalaya

Lolita M. Shangpliang

Introduction

The state of Meghalaya is bounded by Bangladesh on the south and south-west and by Assam on the north and the east. The state shares 443 km. long international border with Bangladesh. Meghalaya is, administratively divided into eleven districts and thirty-nine C.D. Blocks of which all the districts, except two, i.e. districts of East Garo Hills and Ri-Bhoi, share international boundary with Bangladesh. Out of 39 C.D. Blocks 10 have international border with Bangladesh. These Blocks are – Zikzak, Dalu, Baghmara, Rongara, Ranikor, Mawsynram, Shella-Bholaganj, Pynursla, Amlarem and Khleihriat. Demographically more than 27 per cent of the state population (4,63,828) live in these ten border Blocks. The above statistics indicate the enormity of area and population of Meghalaya which are directly within the socio-economic impact of Bangladesh border.

Agriculture is the principal occupation of the border people. The main crops grown by them are areca nut, betel leaves, bay leaves, and orange. Rearing of poultry and pigs by households is another common activity in these villages. Growing of broomstick in large tracks of steep hills is another feature of the border villages. In recent years some households in Pynursla have started coffee plantation too. In Shella-Bholaganj bee keeping is another important economic activity of many families.

This work makes an attempt to trace the history of border trade between the districts of Sylhet in Bangladesh and Khasi Hills in Meghalaya and to focus on the impact of Balat border Haat, an important border market located in Mawsynram Development Block, on the socio-economic life of the border people of Meghalaya

Historical Background

The trade relations between Bengal and Surma valley of Assam through Sylhet existed ever since Mughal period. British government records reveal that the Khasi and Jaintia Hills district had a trade relations with Sylhet district of Bengal when the British had first intercourse with them. This is evident from the fact that – "In 1774 an expedition was sent against the Raja of Jaintia who was accused of obstructing trade in Surma."[1] During pre-British time transactions were mostly carried on barter system and also through payment by cowrees (sea shells) although some Khasi Syiems are said to have struck coins for internal transaction. Immediately after their arrival the British saw the potentiality of natural resources of Khasi and Jaintia Hills and realized the importance of trade with Bengal. Accordingly they reorganized the markets on a better footing and introduced standard weights and measures in place of prevailing crude barter system. The Government struck coins and converted barter system into monetary one. The following report of Col. Lister, the Political Agent in-charge of Khasi and Jaintia Hills, gives us an account of the volume of trade in respect of agricultural, horticultural and mineral products that existed between the Khasis and their counterpart in Bengal.[2]

> "Of horticultural produce, oranges take the lead. The plantations are situated at the foot of the southern face of the hills, and reach to an elevation of about 1000 to 1200 feet. The export is about ₹ 400 to 500 lakh a year and is still increasing". . . .
>
> "Potatoes are next in value, they are cultivated on the table land and on slopes. . . . The amount cultivated is estimated at 30,000 maunds. . . . the gambling character induces them frequently to take their crop to Calcutta in the hope of a large return and frequently to their complete ruin."
>
> "Coal is next in importance, but the difficulty and expense of carriage prevent its being of much commercial importance."
>
> "Iron is found in the interior of the Khasia range. . . . The greater part of it is being exported to Assam and Sylhet".
>
> "The import of the district (Khasi Hills) comprise rice, salt, tobacco, dried fish, oil, corals, brass and copper utensils of all descriptions, cattle and goats".

As the trade further expanded the nature and volume of trade underwent significant changes in the late 18th and early 19th century. The British merchants started lime trade which became a monopoly business for many years to come. Together with lime and coal trade other agricultural produce, oranges, betel nuts, pan leaves produced by the Khasi villagers found easy outlet into Bengal. Thus the border tribal people in pre-colonial times earned their livelihood by selling agricultural produce most of which being highly perishable like oranges, betel-leaves, vegetables, pineapples etc. Other important agricultural products were bay leaves, pepper, yam, areca nuts etc. The hills are also famous for forest products namely timber, bamboo and lac. Mineral products like limestone and coal and iron used to be exported to Sylhet in large quantities after the British came to Khasi Hills. Iron smelting and manufacturing agricultural implements and household tools like daggers knives etc. Kodalis was another source of livelihood of the Khasis. These iron tools and implements used to be sold in the border haats and they had good demand in the plains districts. The cottage industry of iron smelting and manufacture of implement had reached a high degree of proficiency even before the coming of the British.

Trade and commerce were important economic activities of the pre-colonial Khasis which were carried out on both internally and externally. They took the inland products to the border markets and exported them outside the State, and at the same time, they bought foreign products and sold them at internal markets located within Khasi Hills. External trade was chiefly conducted through the markets on the southern foothills across Sylhet district and the chief articles of exports were as mentioned above. The import items consisted of rice, salt, factory goods and other necessities which the hills lacked. To enhance and promote trade, the Khasi chiefs established markets or *haats*, on a regular basis and saw to their management. The principal markets at the foothills on the Sylhet side are Bholaganj, Chhatak, Lakhat, Jaintiapur, Jaflong, Pharal bazaar, Maodon, Sonapur, Molagul, Pandua and Lenjut in Sylhet. Many of them are situated within Sylhet boundary, but they are all frequented by the Khasi and Synteng traders who spend their whole time in passing from one market village to another.[3]

Usually the Khasi markets were held weekly. But there were some that were held bi-weekly such as the Iewduh market at Shillong, the Shella, Mawsynram, Tyllap Markets etc. These markets exist till today in the inland areas.

B.C. Allen, in the *Gazetteer of the Khasi and Jaintia Hills District*, gives an account of the traditional trade as follows:

> "The Khasis are keen traders and considerable portions of the population earn their living by travelling round to the markets held at different places in the districts. The principal exports from the district are lime, potatoes, oranges, betel nut, pan and other foods grains, dried fish, salt, tobacco and hand woven cotton and silk cloth from the plains."[4]

Thus, trade formed an essential part of the economy of the Khasi and Jaintia Hills. The transport system being greatly inconvenient by the existence of hilly terrains commodities were mostly carried by head loads across the hills down to the border haats. Robert Lindsay in his book, *Anecdotes of Indian Life*, wrote:

> "During the few days of my residence at Pondua, I had the uncommon gratification of witnessing a caravan arrive from the interior mountains bringing on their shoulders the produce of the hills consisting of coarsest silk from the confines of China, fruits of various kinds, but the great staple was iron of excellent quality. In descending the mountains was by steps out of the precipice."[5]

Thus bringing their products to the haats for sale and buying the necessities of life every week was the established tradition of the Khasis. But after the partition of India in 1947 the age-old trade with the plains (Sylhet) came to abrupt halt. The borders were sealed and the haats were closed as a result of strained political relations with Pakistan. The hill people suffered a great loss due to lack of market for their perishable goods like fruits and vegetables. Prices of potato and oranges went down beyond the cost of production. The after-effect of the partition brought untold miseries to the people of the area who were the worst sufferers. Due to hilly terrain and absence of road transport facilities the agricultural produces grown in the border areas could not be carried to the mainland marketing centers. As Hamlet Bareh wrote – the staple food of the Khasis such as rice which was obtained from Sylhet during pre-partition period was not available due to the closure of border haats and the food scarcity and near starvations occurred at intervals to the extent that the people had to survive with wild vegetables and roots. The trade of betel nuts and *tezpata* leaf which earlier exceeded the volume of income yielded by potato, was now limited both in production and circulation.[6]

This situation of deadlock gave rise to illegal trade and smuggling of goods, mostly of Chinese products from East Pakistan. Cattle lifting, infiltration of Muslims into India and various types of illegal activities became rampant. Government of India erected barbed wire fencing along the border and stationed Border Security Force to check illegal transactions, but to no avail. After Bangladesh came into being in 1971 and the relations with India having improved an informal trade had commenced with the expectation of reopening of border trade. But the expectation was belied when trade between the two countries was totally stopped and the whole border was sealed off again. A comment in the Kolkata newspaper on Meghalaya ran as follows:

> "The present volume of trade and output is a trickle compared to pre-independence figure, since the economy was oriented towards East Bengal plains and the available riverine facilities. An official dossier records that the prices of produce sank to unbelievably low levels. Worse still, there was no one to lift the produce in the absence of easy transport link with the alternative markets in the rest of India. . . . With the opening of a few vital roads alternative markets were found but transporting the produce to the focal points posed a problem. Transport by head loads was primitive and possible only in the non-monsoon period.[7]

After prolonged negotiation, towards the quarter ending March 1971 a few border *haats* were declared open for transaction in the listed articles only. Before the other developments ensued, the maintenance of relations was upset by the Bangladesh affairs bringing with it influxes of refugees into the state of Meghalaya.[8] The influx of refugees from East Bengal during the Indo-Pak war on 1971 created an unprecedented setback in the border markets in particular and in the whole state of Meghalaya in general. Some steps were taken after the liberation of Bangladesh to normalize trade relations following the signing of a Trade Agreement, but it was suspended later on.[9]

Balat Border Haat

The Balat border haat was finally declared opened on October 23, 2010 at Lalpani, Zeroline, at Balat within the jurisdiction of the Bhowal

Syiemship. The Government of India and the Government of the People's Republic of Bangladesh entered into a Memorandum of Understanding along with the Mode of Operation of Border Haats to promote the well-being of the people dwelling in remote areas across the border of the two countries.

The Haat Balat is situated at a distance of 102 km from Shillong, the capital city of Meghalaya and is linked with all-weather PWD road. It is one of the chief trading and marketing centers in the East Khasi Hills district. During pre-partition days Balat was a flourishing centre of trade and commerce between Sylhet and the southern border of Khasi and Jaintia Hills but after partition in 1947 the volume of transaction has greatly reduced due to termination of bilateral trade relations. This *Haat* provides trading facilities to a large number of surrounding villages, like Trumpleng, Mawpen, Kynrang, Sohsaw, Ranikor and Mawdon, Mawkhan, Lawbah, Ranku, Mawsynram, Syntein, Weiloi etc. falling under Mawsynram Development Block. The commodities produced by these villages are transported down to Balat haat every Tuesday, on which the weekly market sits. The mode of transport is normally on head loads or by bus, or Jeeps wherever roads are available. Nowadays growing of broom-sticks in the steep hill slopes has become a good source of income for the border people.

As directed by the government of India, the Commerce and Industries Department had constituted a *Haat* Management Committee for the Border *Haat* at Balat, East Khasi Hills District. The joint *Haat* Management Committee between India and Bangladesh hold regular meetings in order to lay down the modalities for conducting such trade. The 1st joint Haat Management Committee was held at Zeroline, Balat (Lalpani) on June 8, 2011. The *Haat* Management Committee meeting was led by D.M. Wallang, Chairman and Additional District Magistrate, East Khasi Hills District, Shillong, Meghalaya (India) and by Md. Kamrul Hassan Khan, Chairman & Additional District Magistrate, Sunamganj District of Bangladesh.[10]

Thus the border haat at Balat has since been providing marketing facilities to the border people of both sides and playing a significant role in sustenance of their socio-economic life. The villagers bring to the haat crops like orange, bayleaf, betel nut, betel leaf, jack fruit, pineapple, banana, mango, potato, honey etc. The commodities brought for sale from the other side (Bangladesh) are: Garments, melamine products processed food items, fruit juice, toiletries, cosmetic, plastic products, aluminium products, cookeries etc. The volume of transaction amounts to not less than ₹ 50 lakh per market day. Transactions are carried out both by barter as well as currency. Counter for exchange of foreign currency has been provided at Balat by Governments of both sides for the convenience of the buyers and sellers.

For the Khasis the weekly Haats are not only the venue for buying and selling of goods but also for other socio-cultural activities. Markets are the places where people met and share their good wishes and ideas each other. With the Bangladeshis they learn languages of each other and share their cultures. Among themselves they dispatch important messages from one village to another through market goers. The District Council and Government Purwanas (circulars) and the village Durbar orders are announced in the markets. Nowadays, markets are used as political platforms for electioneering purposes. Thus the weekly haats, besides being the places for buying and selling of goods, are the centers of publicity, propaganda and education.

NOTES

1. R.B. Pemberton, *Report on the Eastern Frontier of British India*, Calcutta, 1835, Sec. 5.
2. A.J.M. Mills, *Report on the Khasi and Jaintia Hills*, 1853, reprinted 1991.
3. W.W. Hunter, *Statistical Account of India*, p. 241.
4. B.C. Allen, *Gazetteer of the Khasi and Jaintia Hills District.*
5. Robert Lindsay, *Anecdotes of Indian Life.*
6. Hamlet Bareh, *The History and Culture of the Khasi People*, Shillong, 1967, p. 466-67.
7. *Statesman*, Calcutta: Dated 2 August, 1971.
8. Hamlet Bareh, *Meghalaya*, Shillong, 1974, pp. 209-10.
9. L.S. Gassah, *op. cit.*, p. 68.
10. Minutes of the 1st Joint Haat Management Committee Balat Border Haat East Khasi Hills District, Shillong, Meghalaya, held on 8th June, 2011 at Balat.

REFERENCES

Allen, B.C. (1980). *Gazetteer of the Khasi and Jaintia Hills District*, Delhi: Gian Publications.

Bareh, Hamlet (1967). *The History and Culture of the Khasi People*, Shillong: Ri Khasi Press.

Bareh, Hamlet (1974). *Meghalaya*, Shillong: North-Eastern India News and Feature Service.

Gassah, L.S. (1998). *Traditional Institutions of Meghalaya: A Case Study of Doloi and his Administration*, New Delhi: Regency Publications.

Lindsay, Robert (1997). *Anecdotes of Indian Life* (ed. David Syiemlieh), Shillong: North-Eastern University Publications.
Minutes of the 1st Joint Haat Management Committee Balat Border Haat East Khasi Hills District, Shillong, Meghalaya, held on 8th June, 2011 at Balat.
Pemberton, R.B. (1835). *Report on the Eastern Frontier of British India*, Calcutta: Sec. 5.
Statesman, Calcutta: Dated 2 August 1971.

SECTION IV

Nature, Culture and the Tribes of North-East India

12

Culture of the Karbis in North-East India

Raima Narzary

Introduction

Due to lack of written materials or documents we have little source about the Karbis. Regarding their history we have to base on 'legend' or 'myth'. The legend of the Karbi had some unique qualities which throw light on their history and their culture. Of course, in the old chronicles and *buranjis* occasional references here and there were made to the people of this race. But from these references also it is very difficult to trace the chronological events of the Karbis. Their folktales and folklores are the only sources from which we can have an idea regarding their early history. The Karbis are a Scheduled Tribe in Assam (North-East India) and represents one of the prominent ethnic groups of the region, distinct in traditions and culture. However, they never call themselves Karbi and sometimes Arleng, which literary means a man. They consider that the term Mikir is given to them by neighbouring people. Ethnically, they are Mongoloid and speak a local dialect belonging to the Tibeto-Burmese particularly Kuki-chin-sub group of languages. The Karbis took refuge in the deep jungles and high hills leaving their hearth and home in the sub-mountain regions.

In traditional Karbi society, every occasion is a ritual and every ritual is an occasion. And there are as many incantations or hymns as there are rituals. This may perhaps sound simplistic, but the presence of a huge number of undocumented practices of a day-to-day Karbi life, seems to confirm this statement.

Marriage

Marriage among the Karbis is strictly enogamous and solemnized through a traditional practice called Adam Asar. Widow remarriage is socially acceptable though extra-marital cases are not uncommon among the Karbis. *Me kehang* is a case of extra-marital affairs, which often leads to public execution or capital punishment. Although monogamy is the prevailing practice, there is no to polygamy and the case of polygamy are very rare. Widow remarriage is allowed by the Karbi society and the unmarried younger brother is required to marry the widow of his elder brother. But under no circumstances, the elder brother is allowed to marry the widow of his younger brother.

Cross-cousin marriage is a preferential one. Marriage by negotiation and marriage by selection of life partners are prevalent among the Karbis. In case of marriage by negotiation, the consent of the girl is a must since a Karbi damsel cannot be forcibly married to a boy of her disliking. Marriage by capture is also found to be present to a limited extent and in this case the formal marriage has to be solemnized after the birth of the first child at a convenient date.

Divorce, which is rather rare in the Karbi society, can be obtained through the approval of the village council and the council gives its approval only when it finds that the separation between the husband and wife is absolutely essential. Like other tribal societies, the Karbi do not have the system of bride price. After marriage the wife continues to use the surname of her father. Part and parcel of Karbi culture – Bong Gourd shell is essentially an integral part of Karbi culture and regularly used during Adam-Asar or marriage and worships. Gourd shell is an indespensible part of Karbi culture and till today it is used in the society.

Family Structure

The Karbi follow the patriarchal system of family structure and as such the father is the head of the family and authority. The line of descent is following through the male member only. A Karbi family, therefore, is a unitary one. Joint family system is also still prevalent to a very limited extent. At the death of the father his sons inherit the moveable as well as immovable properties and not by the daughters. If the father dies without any male issue, his property is inherited by the nearest male relatives of his clan. A childless couple can adopt a son which must belong to the clan of his foster father and at his death, the adopted son will inherit the property.

Social Life

The settlement pattern of the Karbis is in the form of a village. The

Karbis like other hill tribes have a tendency to live on the hill tops. The villages are not only smaller in size, but scattered too. In the plain portion of the Karbi Anglong district where the Karbi people practise permanent cultivation and where the villages were quite strong, the villages were found to be stationary. But in the interior area of the district shifting cultivation is practised, shifting of village site is still in practice.

The post of the village headman whether that of the revenue village or of the hamlet is a very coveted and prestigious one. In the performance of every socio-religious rite or festival, the headman has to be honoured first. Moreover, the village or the hamlet is also known after him. If the Karbi people living in a particular village think that their village is a haunted place frequented by ghost or evil spirits, they shift their village to a new site very soon to get rid of the ghost or the evil spirits. The Karbis who practice *jhuming* or shifting cultivation very often shift their villages to new *jhum* sites in order to stop this habit of shifting and also to wean away people from shifting cultivation.

Houses of the Karbis

A typical Karbi hut is neither too small nor too big. It is built on bamboo platform using timber posts for superstructure. The platform is several feet high above the ground. For roofing purpose thatch is used, the houses has two verandas one at the front and other at the rear. The house is divided into two parts lengthwise. The front part or room with a hearth at the centre is called *kam* or guest room. A wooden or bamboo ladder is used as an approach to the front veranda. In the plain areas of course the traditional housing pattern has more or less been abandoned by the Karbis. Instead of having raised bamboo platforms, the houses are constructed on grounds. Building materials required for construction of the houses, whether traditional or modern, are procured from the nearby forests.

Birth or Death

Incarnation they considered almighty as "Arnam Sansar Recho". The birth of a child whether male or female is an occasion for joy in the family in the sense that a person died earlier in the family is supposed to be reborn in the same family. In other words, the Karbi believe in rebirth as human beings and not in the form of animals, birds or other creatures.

Hence, for a new born child the name giving ceremony where the maternal uncle of the child takes the leading part is the most important one. If the baby is a female one he will utter the names of all the dead females one by one associating a name with a few drops of wine. The name that will be associated with the last drop will automatically be the name of the female child. Similar procedures of associating each drop of rice-beer with the names of each dead male in the family are followed in case of name giving ceremony of a male child and the name associated with the last drop will determine the name of the newly born male baby.

Death in the family, on the other hand is considered to be the most tragic incident in the family. For the cremation of the deceased, the Karbis observe some specific rules and regulations. Although they perform death ceremony at the time of the cremation of the deceased, they also perform the death ceremony called *chomangkan* at the later death for the eternal peace of the deceased. The performance of the death ceremony is must for the Karbis irrespective of social status and economic position. It is the most elaborate and expensive socio-religious ceremony of the Karbis which continues for four days and four nights non-stop.

Religion

From the point of view of religion, the Karbis can be regarded as animists. They believe in the immortality of the soul, life hereafter each disease is associated with a presiding deity. Hemphu and Mukrang are two benevolent household Gods. For the appeasement of the deities, the Karbis observe many religious rituals throughout the year where the sacrifice of pigs and birds and use of rice beer are indispensable. They also believe in witchcraft and black magic. A smaller section of the Karbis has embraced Christianity.

Economy

Agriculture is the primary occupation of the Karbis, in the hilly areas they practice shifting cultivation besides paddy they also cultivate other staple food as well as vegetables and mixed crops. Shifting cultivation is known as *jhum* cultivation in the North-East India. The *jhum* fields are protected from wild animals, stray cattle and birds by constant watching till the harvest is completed. After cultivating the *jhum* for two years they select another plot for cultivation. Handloom industries are very much famous among all the Karbi families. Spinning and weaving is unthinkable in the Karbi society. All the Karbi women are expert weaver in the *jhum* cultivation they cultivate cotton also.

Thus, the Karbi have an economic system in which they produce what they actually need leaving either no surplus or little surplus for sale. Their economic condition is, therefore, far from satisfactory and this is further aggravated by the fact that a considerable portion of paddy produced by each family is used for brewing rice-beer which is required for the performance of all social festivals and religious rites, for domestic consumption and for the entertainment of guests as well.

Festivals

Among the festivals observed by the Karbis, mention may be made of the Rongkher and Hacha. While former is socio-religious and latter is a social one. Rongkher is performed at the beginning of the New Year by propitiating the different gods and goddesses for the well-being of the entire village. The deities are worshipped by all the elderly male members of the village so that with their blessings the people of the family could have a good harvest. The Karbi perform another festival called Hacha after the harvest. This is rather a merry-making festival where all the members of community eat, sing and dance to one's heart content are the most striking features of this festival.

Dresses

The Karbis have their traditional dresses which are artistically designed. These dresses are woven at their family looms. There are separate dresses for men and women. The aged men use an artistically designed shirt called *soinangpo* and the shirt used by the young men is *saihothor*. The men use a loin cloth called *Rikong*. *Rikong* is found to be used specially by married and aged persons. The use of turban called *poho* is also fast disappearing. The Karbi women and girls generally use *pinilamflak*, a piece of cloth tied around the waist like a mekhela. A piece of artistic cloth is used by them to cover the upper part of their body and it is called *jar-ek*. A very highly artistic waistband called *wampo* is also used by every women and girl. During the performance of *chomangkon*, young girls use a special scarf called *Dokherso*. The Karbi women and girls are fond of their traditional dresses and they have been using them even in the face of a strong competition of the modern trends.

Dance and Music

Dance and music play an important role in the life of the Karbi society. All Karbi youths, both boys and girls know dancing. Various types of dances are performed by the youths during the performance of *chomangkan*, the ceremony and the religious festivals. *Hacha kekan*, the dance performed at the harvesting festival is very lively and eye-catching. Their traditional songs either folk or religious are sung by the expert. During the time of death cremation rites only a professional weeper called *uchepi* is allowed to sing a melancholy song called *Sarhe*. They have a Karbi version of the *Ramayana*, but the most interesting part of this epic is that it is unwritten and it has been handed down from one generation to another orally.

The Karbi have limited number of musical instruments. A big drum called *cheng* is their main musical instrument. It is generally played by master drummer called *Duihudi*. Small drum is called *chengbruk*. In some of their dances, they use war shield made of rhinocero's skin called *song*.

Food Habits

Like other tribes of Assam, the staple food of the Karbis is rice. Other cereals like *dal* are occasionally taken with rice. Vegetables grown in their *jhums*, wild roots and tubers and eatable leaves collected from their nearby forests are also taken by them along with rice. Curry prepared with dry fish is their favourite delicacy. Rice is taken twice daily in the morning hours and in the evening. Pork and chicken are also very much liked by them. The most favourite beverage of the Karbis is rice-beer which is brewed by every family at home. It is a prerequisite for the celebration of every festival whether social, religious or socio-religious. Black tea without sugar and milk is taken by them occasionally. They generally do not drink milk, but those who are living in the urban, semi-urban and growth centres, are in the habit of drinking tea with sugar and milk.

Conclusion

During the recent years, the Karbi society has undergone transformation and changes. The grip of inertia in which this society was found in the past has now been slackened. From the point of view the Karbi had very rich culture containing strong folk-lores and folk-songs. Now, the Karbi women have also been coming up to accept the challenges of translation.

REFERENCES

Articles and Journals.
Karbi Studies, Vols. 1, 2, and 3.
Tribes of Assam, Parts-1, 2.

13

An Angami Perception of Nature: A Beckoning for Inter-cultural Dialogue and Criticism

Dominic Meyieho

Introduction

In the contemporary culture, the Human relationship with the Nature is that of *subject* and *object*, i.e. the Human as mistress/master and the Nature as resource. Contrary to this view, this chapter presents the relationship between the Human and the Nature from an Angami perspective, in whose culture the differentiation between them is minimal. The Angamis perceive themselves as part of the Nature and play with its seasons. In the light of the above two perceptions, the chapter invites a cross-cultural dialogue and mutual criticism.

Contemporary Scientific Perception of the Nature

To the world of science and technology, the Cosmos exists only for the Human use, experiment and to gratify a proliferation of his/her needs. Anything and everything in the Cosmos to be meaningful and have value bears the label of humanistic. A key figure in developing a humanistic and scientific approach to the Cosmos was Francis Bacon. The humanistic approach to the Cosmos perceives the Cosmos as existing for the Human experiment. It perceives the calling of the Human as to put the Cosmos into test and perfect the imperfect Cosmos. The humanistic and scientific approach to the Cosmos is anthropocentric and not cosmocentric. Therefore, the whole endeavour of humanistic approach to the Cosmos is to build the Human 'estate' by experimenting and putting the Cosmos into test (Hawkin, 2004: 29). By doing so, the

Cosmos is reduced to a quantitative order, and the Human with his/her reason assumed the measure of all things.

To the scientific approach, the Cosmos is just not given but it is as a result of evolution and continues evolve. It perceives the Cosmos in the calculative dimension of 'cause-effect', and thinks that the whole events of the Cosmos can be understood and be predicted. In this calculative dimension of 'cause-effect', the Cosmos is not a *subject* and *good*, but instead it is an *object* and is perceived as *goods* (commodity) to be bought and sold. Hence, the Cosmos is perceived as though it exist solely for the gratification and proliferation of the Human needs.

In the gratification and proliferation of the Human needs, the Cosmic dimension is relegated to mere 'matter' or 'raw-material' of utilitarian significance. The perception of the Cosmic as 'matter' suggests that in no way the Cosmos is divine or suffused with divine, but it lies open subjected to the test of the Human reasoning, i.e. to improve and build his/her 'estate'. The Human civilization, progress and development are measured by the increase rate of the Human wants and the degree he/she make use of the Cosmos. But in making use of the Cosmos to meet its wants, it is perceived as an enemy to be defeated, conquered, subdued and controlled, i.e. the Human mastery over the happenings of the Cosmos. Therefore, the proliferation of the Human needs is by exploiting the Cosmos (Raj, 1998: 142).

Even Christianity, one of the major religions of the world perceives the Cosmos as subjected to the Human. It seems as though the Cosmos is created for the Human pleasure and enjoyment, and thought the Human to be the master and lord of creation:

> Be fruitful and multiply, and fill the earth and subdue it; and have dominion over the fish of the sea and over the birds of the air and over every living thing that moves upon the earth (*Genesis,* 1. 28b).

It appears that the whole creation is for the Human to enable him/her to work out his/her salvation or liberation through the use of the created Cosmos. But perhaps, it is theologically sinful and philosophically naïve to speak of the Human only as in need of liberation. In our times today, not only the Human but also the created Cosmos needs liberation so that it is not destroyed but cultivated. Thich Nhat Hanh (2008) says, "the Earth is ailing, society is ailing, and there is so much despair and violence" (p. 12). "Today, more and more attempts need to be made to regain the cosmic dimension through ecological measures and programmes" (Raj, 1998: 143).

The advancement of humanistic science and religion – Divine as "God – the Father", tribal belief of sacredness of the Cosmos and its belief system is relegated as rubbish, primitive, uncivilized and uneducated. The humanistic science and religion perceive tribal as lacking human dimension and has nothing to contribute to the well-being and enrichment of humanity, but only at the receiving end. Today, with the "God-Father" concept of religion, the mystery and sacredness of Nature is rapidly violated. There is no sacred time and situation to plant, nurture and harvest the Nature but anytime is permitted in the humanistic religion "God-Father" who is omnipresent, omnipotent and omniscient. Today, this concept of "God-Father" religion of omnipotent, omniscient and omnipresent is destroying Nature without letting Nature to rest in its secret to re-create and pro-create.

An *Angami* Perception of the Nature

The *Angami* life is a play with the different seasons of the Nature. The rule of this play for the *Angami* is *Nanyü*, which he/she performs for peaceful, mutual co-existence with Nature and its spirits. The indigenous belief of the *Angami* holds that all things are animate and all phenomena are permeated by spirit. This belief of the *Angami* is expressed culturally. Most tribal religion and especially the *Angami* do not speak of religion as separate from the day to day living but a way of life that conducts and directs one's life – it is a myth lived out. The *Angami* called this way of life which conducts his/her life as *Nanyü*. *Nanyü* in its limited sense can be called as religion; however, *Nanyü* is beyond religion. *Nanyü* is the cultural expression of the essence of *Angami* way of life and living.

Nanyü is the custodian of *Angami* life, i.e. it is a skim of law, value, and guideline for an individual and society to conduct life. *Nanyü* is more for a relational purpose – a peaceful co-existence. The *Angami* does *Nanyü* to everything that forms part of his/her life. It is *Nanyü* that decides all his/her activities. The *Angami* being agriculturists has lots of *Nanyü* for Nature and natural forces for mutual benefits and peaceful co-existence. The *Angami* believes that he/she is part of the Nature in as much as the Nature is part of him/her, and he/she does not exploit Nature – his/her own body.

The seasonal crops, which come as the Nature's bounty, are not consumed without *Nanyü*. The *Liedepfü* (the first harvester) first tastes the first fruit of Nature and gives thanks and pronounces a blessing on its people that those who eat may be strong and healthy, and no evil may befall them. The *Nanyü* to the seasonal crops is also an acknowledgement of the Human role in making the Nature productive. Hence, for every initiation and harvest of the seasonal crops *Nanyü* observed. The observation of *Nanyü* is the human consciousness of himself/herself and the Nature. The *Angami* sees the Nature as part of himself/herself and at no point of time the *Angami* subdues the Nature, but appeases Her for mutual benefit and enhancement. To the *Angami*, the Nature is its home – the *oikos*, the habitat of the Human and entails the mystery to be unfolded. In Nature, the Divine is subsumed and hence it is not merely "natural" but "sacred". To quote Panikkar (1998):

> The entire World is Man's habitat; he lives on and cultivates the Earth. He has no "sense" of Nature, for he is part of it. He does not feel the need to contemplate Nature, since he himself belongs to it. He hunts, fishes and toils upon the Earth . . . . He is neither a spectator nor an actor on Earth, but its "natural" product. He is thus sacred, for the entire universe is sacred, and he is a part of the whole (p. 25).

The appeasement and its ritual to Nature is a worship of the Nature, by which the Human in turn worship itself who is part of Nature. Therefore, the appeasements and its rituals of the *Angami* has not found any concrete manifestation either in the form of structure like the Church or Temple or *Gurudwara* or Mosque or *Sangha* etc., or sacred text, but conducting life in the ways of Nature – a way of life. Thus, the *Angami Naga* being an agrarian tribe, all its festivals and works are related to the different seasons of the Nature. The commencement of any work or season of Nature is marked by celebration and the observation of *penyie* (a day of obligation and abstinence from manual work).

The *Angami* by nature is religious. He holds that there is a cosmic soul, and everything in it has a soul and strife his best to live in communion and harmony for mutual benefit and enhancement. The *Angami* lives on the lap of Nature and believes that the slightest disturbance on Nature affects him/her. For the *Angami*, Nature is alive, a living being with its own origin, growth, development and dynamism (Jayard, 2005: 23). Nature is a living spirit, it is *anima mundi* – a living organism, and that it is not a corpse enlivened by a soul (Panikkar, 1998: 139). The Nature is the teacher and guide whom the *Angami* follows. She is his/her sustainer, provider and She is his/her way of life and living. She is not a deceiver in the sense that all the activities of the *Angami* is centred and ordered in and after the Mother Nature. Thus, the

Angami appeases Her with rituals and sacrifices for mutual co-existence and goodness. He/she loves and owes his/her respect to Her – the Mother Nature.

The *Angami* celebrates six major feasts (*thenyi*) in a year. These feasts (*thenyi*) are in relation to different agricultural seasons. The *Angami* being an agriculturalist celebrates the different seasons of the year along with his/her agricultural activities, which are based and ordered in the different seasons of the Nature. Perhaps, at the closer look we may find, the different feasts (*thenyi*) of the *Angami* is a celebration of the Nature happenings along with his/her human endeavours and innovations in toiling and cultivating Nature and making her productive for mutual benefits.

Tsiakranyi (Feast of sowing) is the first feast of the *Angami Naga*. With this feast of *Tsiakranyi* begins the sowing of seeds. It is a taboo for the *Angamis* to sow before the feast of *Tsiakranyi*. Soon after this feast, the *Angami Naga* prepares his/her fields for the sowing of crops such as maize, jobstear, millet etc. The sowing of seeds is first initiated by *Tsiakrü-u* (the first sower). *Tsiakranyi* on the whole is a feast for the commencement of *Jhum* cultivation.

In the fifth month of the *Angami* calendar *Kerünyi* is celebrated. With this feast commences the transplantation in the terrace field or wet rice cultivation. However, much like the sowing of seeds by the *Tsiakrü-u* (the first sower), even for transplantation there is the first initiator. It is only after the *tekhu sekhrü* (initiation of transplantation) the community is permitted to do so. We may call this feast of *Kerünyi* as the monsoon feast.

After the millet harvest, a feast is celebrated known as *Chünyi* (Millet feast). *Chünyi* is a seven days celebration, which is celebrated with much gaiety and merrymaking.

When the wet rice cultivation begins to ripen the *Angami Naga* prepares for the harvest by clearing the foot-path to the fields. During this period, a feast known as *Chadanyi* (Feast of road/way making) is celebrated. This feast of *Chadanyi* is celebrated by the Kohima village and not by the Southern *Angami* villages. In the place of *Chadanyi*, the Southern *Angami* celebrates *Kezinyi*.

To prepare for the harvest, the Southern *Angami Naga* celebrates a feast known as *Kezinyi* (Feast of Advent). In this feast of *Kezinyi*, a three days *penyie* is observed, after which the *Angami Naga* makes baskets, mats and other items needed for the harvest.

After the harvest of the wet rice cultivation the *Angami Naga* celebrates *Terhünyi* (Feast of rice or harvest feast). This feast marks the longest observation of *penyie* to enjoy the fruit of the harvest. On this feast of *Terhünyi*, the *Angami Naga* dressed in traditional ceremonial attires and engages in merry-makings.

Soon after *Terhünyi* comes the feast of *Sikrenyi*, the biggest of the festival of the *Angami Naga* tribe. Most of the Southern *Angami* celebrates *Sikrenyi* in the months of November-December. The Kohima groups celebrate it in the month of February. Unlike other feasts there is hardly any gap between the feasts of *Terhünyi* and *Sikrenyi*. *Sikrenyi* is a feast of purification and a new beginning. Therefore, it is also known as *Phusanyi* (Feast of Purification).

The popular understanding of the feast of *Sikrenyi* is that it is a feast of the male folk or a feast of male chauvinism. During the *Sikrenyi* it is a taboo for women folk to fetch water and sleep with their husband as they are considered physically unclean, as the feast of *Sikrenyi* is all about a new beginning and sanctification or purification, therefore, no defilement from any quarter. However, if we see it from an ecological perspective, we may say, the Nature and Woman who is engaged with Mother Nature all the year round is given rest, to nurture and to re-fresh herself until the sowing season arrives.

Cross-Cultural Dialogue

Nature is sacred and secretive, and she is ever unfolding herself. The hills, lakes, valleys, rivers, seas, oceans, forests, deserts, beasts, plants etc., are all part of the sacred family of the Cosmos. The life of the *Naga* tribe in general is a life with the Nature. A *Naga* start his/her life in a cradle of bamboo and ends in a coffin of bamboo. An old man of Khari village aptly and fittingly remarked: "I construct my house with bamboo, use bamboo utensils and equipment in the bamboo house, burn dried bamboo as fuel, use bamboo torches and eat bamboo pickles (*The Arts and Crafts of Nagaland*, 1968: 74)". Until very recently wooden platters, dishes and all kinds of spoons for serving and taking food are made at home out of Nature's bounty. There is a popular large four-legged platter from which the couples usually dined. Not only are the plates of the *Naga* of bare Nature but other utensils as well. Receptacles for storing, carrying water and rice-beer are made of gourd grown. Gourds are also cut to form ladles, the broad end being used for ladling, while the neck forms the handle (Ganguli, 1993: 21-22).

The elephant, tiger and hornbill are the symbols of valour, the *mithun* (Bison) is the symbol of wealth in the general *Naga* society. Some representation of these animals or skulls appears or is used as decorative motif. In some tribe, *mithun* is used as currency to settle marriage or pay fine and penalty (*The Arts and Crafts of Nagaland*, 1968: 66).

It is a general belief among the *Angamis* that to harm Nature is but to harm oneself. Therefore, exploitation of the Nature is foreign to the *Angami* tribe. But on the other hand, the *Angami* revere and respect Her inasmuch as he/she respects and rever his elders and ancestors who form part of the Nature. The *Angami* extend his filial piety and love to the Nature as his dead becomes part of her. The Nature is, thus, alive to the *Angami*, therefore, he/she is invited to love, respect and save Nature, which in turn will respect, love and save him. When the *Angami Naga* cuts his/her livestock for food he/she would bless it by pronouncing something like "may you go the right path and may your flesh strengthen me and not weaken me in any form". The same sort of blessing is invoked even on plants, stones and trees, when he makes use of them admitting the life and spirit on every being both animate and inanimate beings.

> You never enjoy the world aright till the sea itself floweth in your veins, till you are clothed with the heavens and crowned with the stars; and perceived yourself to be the sole heir of the whole world, and more then so, because men are in it who are everyone sole heir as well as you (Panikkar, 1977: 86).

To the *Angami* the Earth (land) is his/her Mother and the Sky is his/her Father. Leaving the Mother land will be the last thing an *Angami* will do. The Earth is considered as his/her birth right as She gives his/her identity and dignity. Apart from the land (his/her birth right) the *Angami* shall be orphaned and shall be a wanderer. The *Angami* cannot live without the Earth just as the Human cannot live without a body. To the *Angami*, the Human could be said to be the eldest son of a widowed mother – the Earth. The *Angami* works on the Earth to enjoy Her blessings; She is his/her home, family, body and Mother. Therefore, the *Angami* commends his/her dead into his/her gentle Mother and she readily receives Her children in all gentleness soon to be part of Her.

A dying *Angami* man blesses his children with land. Land symbolizes identity and dignity among the *Nagas* in general and the *Angami Naga* in particular. It is seldom to find a person without land, at least the community's land. Allocation of Land by the head of the family to his children is his final and the best of all his blessings. Seldom, an *Angami* man divides his Land when he is young. An *Angami* blesses his children with land because She is the dispenser of every good things and blessings. He entrusts the piece of land to his children that once sustained the family and held them together. To part with one's land for the *Angami* is to part with the blessings of his ancestors. Thus, even after the *Angami* has sold and given his land to government or private institutions, he likes to be identified and be known as the landowner, for to be without land is to be without identity and without blessings.

The *Angami* dwells deep in the mystery of the Nature rather than mastering Nature. Not every time and anytime plants and animals could be nurtured and harvested at will by him/her but a specific time allotted for it. Thus, the mystery which entails in the Nature is ever-new, fresh and worth living it out. For instance, Cactus is not planted as and when he/she so desires but a specified period is allotted for planting it.[1] Likewise, so many other plants and animals have their own specified period to be planted, nurtured and be harvested, and not anytime and every time is permitted.

There is an increasing need and necessity to realize and consider the feminine aspect of what tribal is and speak of the God-Head as feminine – *Ukepenouphfü*, so as there could be acceptance, conceiving (a purgative of female), growth, development and delivery that is the giving for the being of Nature and its wholeness. Potent with producing and production the Earth is referred as female and most of all the Mother of human kind, which owes the Human respect and love. However, with the exploitation of the Earth for pleasure knocking at Man's door, the respect for women is fast ebbing out even in the *Angami* society, which once treasured its women as initiator and custodian of its society.

Discussion and Concluding Remarks

The ecological predicament, the Human faces so painfully and crucially these days, is the most obvious way in which the crisis of historical consciousness manifests itself. The high-handedness of *logos* and progressive thinking which is equated with 'scientific' thinking is an onslaught to the mystery of the Nature with its masculine *logos*. Mastery over mystery with its goal to 'create' heaven on earth is monitored by proficiency and predictability, production and profit, and automation and uniformity. Today, with the *mantra* of mastery instead of creating heaven on earth it has lit the fire of hell on earth with the ozone layers getting burnt off and iceberg melting causing the oceans and seas to rise, global warming and most of all discovering the limitation and finitude of the Earth's resources.

Therefore, today, the Human to save the world needs a heart, a heart that understands, believes and has faith even when reason goes otherwise. The Human need to be part of the mystery that is in the process of the unfolding rather than all finished, complete and perfect. We need to be involved in the game of 'win-win' and not profit motivated and predictability but 'giving' and 'taking' and *vice versa*. Perhaps, to save our environment and ecological disaster from further degradation we need to visit those traditional sources – the cultures that promote the Nature and see its well-being in the health of Nature.

Mere academic discussions, seminars, conferences on ecological disaster and crisis, although, needed yet are perhaps not enough to secure the health of the Nature. What the Human need is to go back to the Nature (*Cosmos*), his/her true home – *oikos* – his/her dwelling place. There is an increasing need to realize the sacredness and mystery of the Cosmos. The Human need to once again dwell in the sacred mystery of clouds that will send showers of blessings on his/her exhausted and dried up Cosmos by its gradual unfolding. The Human need the showers of blessing (unfolding of Nature's sacred mystery) to fall afresh on him and cool his/her Cosmos and bring forth life – life in abundance from its abundance. This is possible by re-visiting those cultures who perceive themselves as part of the Nature or in whose culture the Nature and the Human is not seen as Object and Subject. Thus, we need an orientation for intercultural dialogue and mutual criticism for mutual learning.

NOTE

1. The belief associated with plantation of cactus in the off-season is natural calamities in the form of heavy storm and rain.

REFERENCES

Ganguli, M. (1993). *Naga Art*, New Delhi: Oxford and IBH Publishing Co. Pvt. Ltd.

Hanh, T.N. (October 2, 2008). "Only You Can Uproot the Terror Menace", *The Times of India*, Chennai, p. 12.

Hawkin, D.J. (2004). (ed.). *The Twenty-First Century Confronts Its Gods: Globalization, Technology and War*, Albany: State University of New York Press.

Jayard, S. (February, 2005). "Nature will not Abandon Us", *The New Leader*, p. 23.

Panikkar, R. (1977). "Collegiate Fragmenta: For an Integration of Reality", F.A. Eigo (ed.), *From Alienation to At-Oneness: Proceedings of the Theology Institute of Villanova.*

Panikkar, R. (1998). *The Cosmotheandric Experience: Emerging Religious Consciousness*, Scott Eastham (ed.) Delhi: Motilal Banarasidass Publishers Private Limited.

Raj, L.A.S. (1998). *A New Hermeneutic of Reality: Raimon Panikkar's Cosmotheandric Vision*, Berlin: Peter Lang.

The Arts and Crafts of Nagaland (1968). Kohima: Naga Institute of Culture, Government of Nagaland.

14

Society and the Environment: Nature in the Life of the Khasis

Kiran Barua Nongrum

Introduction

Planet Earth in the solar system is a unique body with life bearing elemental ingredients potential enough for sustaining life from minute living being single cell to gigantic beings since ages. Diverse in its tectonic formation, distribution of water, meteorological conditions, hills and mountainous terrain vegetation system. Living species and its dynamic mechanism have created interest in human world as Environment.

Environment is the nuclei of life. It not only directs but determines the growth, existence and development of human beings and all their activities.

Human beings are just as other living organisms sharing space with other species in an environment and are mutually dependent on each other.

Human knowledge, beliefs and social habits are a by-product of the interaction between man and nature. Human societies since its inception have been nurtured by nature which is a genesis for theoretical understanding between Man and Environment.

The North-East India constitutes the eastern frontier of Indian Nation. Nature has endowed it with sufficient natural resources [such as river system, fertile soil, fossil fuels, forests materials, flora and fauna] within the beautiful Eastern Himalayan chain of hills.

The people in this region are indigenous tribal communities respectively in all eight states namely the Nagas, Mizos, Khasis, Jaintias,

Garo, Reangs, Khacharis, Bodos Monpas, Darlongs, Apatanis, Adis, Wanchos, Tiwas, Tangas, Tipras, Daplas and many more are the habitats of this region for whom life rotates within the orbit of its physical environmental bliss and its challenges. They also share a close relationship with the environment which has remarkably been generated through the process of labour. Using cognizance for survivality, stated by Karl Marx, Labour is primarily "a process in which both Man and Nature participate, and in which man of his own accord starts, regulates, and controls the material re-actions between himself and Nature". It is indeed the process of labour which is the basis for the relationship between man and his natural environment have predetermined man's transition to social life, it is thus labour as a prime factor that has formed the elementary roots for the existence and development of society, human beings have always organized social structures and meanings to avail the benefit of environment and have exercised within social structure.

This chapter on the "Khasis of Meghalaya" is an Endeavour to understand how human social structure and meanings teamed up with labour have marked human evolution and development close with nature and environment.

The Khasis of Meghalaya in North-East India have traditional Khasi Social Organization in which they have social division comprising of Pnars/Jaintias, Khyrims, Bhois, Wars and Lyngngams. Their territorial division is known as Hima and the Chiefs who ruled the Hima are called the Syiems. The Khasis are tribe who adhere to matrilineal system originating from the Ka Iawbei (the ancestral mother), U Thaw Lang (the ancestral Father), U Suidnia (the ancestral uncle), U Knih (the maternal uncle), Ka Khadduh (the youngest daughter).Traditional rights and ancestral property is inherited by the youngest daughter 'Ka Khadduh' who is the custodian of the ancestral property. Khasi livelihood and means of survival strategy have been fostered through their experience with environment. Traditional lifestyle and social set up is motivated by nature. The Khasis revere the Earth and addresses it as Mei Ram Ew (Mother Earth).

An insight as to why earth is addressed and personified as mother nature rests on Khasi legendary conception about cosmic entity, where Ka Ram Ew (mother earth) as her husband Basa were blessed with five children Ka Sngi (sun), U Bnai (Moon), Ka Um (water), Ka Lyer (Air) and Ka Ding (Fire). Their combined effort promoted the growth of plants, fruits, flowers, trees which filled mother earth with greenery and bestowed happiness on her life

Such legendary stories have more significance in fostering a repository content in human and social context which without much difficulty has transmitted why mother earth is held at the apex in regard to nature. The personification of comic bodies in the legendary mode has been transcendental into the world of human, which is reflected by man with utmost respect and adulation for life without any of these comic elements life is impossible – they are life generating bodies and the legends passes with a message to love their land whom they call it Ka Ri U Blei (land of god) for such reasons a Khasi loves and respects the environment in relation to god and mother earth.

Their simple traditional knowledge have maintained best how and how much to use material resources for a purposive need and its genesis rests on traditional beliefs, ethics, myths and folklores as divine and sacred origin where their socio-economic livelihood depends much on Nature, in its various forms – forests, rivers, hills and trees which are also the abode of spirits and deities perceived since time immemorial through oral-tradition, as fables, legends and myths. Addressed as 'Khana Tang bad Puriskam In most of the fables and myths like the Kwai Tympew Shun Bad U Duma Sla (betel nut, betel leave, lime and tobacco) U Sier Lapalang (the reindeer), U Klew Bad Ka Sngi (the peacock and the sun), Ka Likai (The waterfalls named after Likai), Ka Pa Syntiew (the story of a girl by the name of Pasyntiew), Ding Iei (The High hill), Lum Sohpet Bneng (The navel hill which is believed to have connected the heaven and earth), U Symper (Rocky Eminice), U Thlen (man eating serpent), U Kopli (the Kopli river). In all these myths, story is build upon natural phenomena – animals, trees, hills, rivers, birds and plants in which the story characters are drawn from the environment, the storylines are well blended with nature so as to elucidate Khasi philosophical thought and culture. Such myths about nature have passed down from generations since time immemorial when means of recordation of thoughts and ideas had not yet evolved, it was only through the art of storytelling by the elders in the society, the role and significance of nature and environment with social message and moral values reached and captivated even the simplest of men for it highlighted values like simplicity, humbleness, love, obedience, compassion, kindness, which socializes human beings to live in harmony with the environment. Learning, knowing and understanding such values requires deep insight and theories with right explanation. They are intensive and loaded with high qualitative analysis. Theories on values are an extensive and deep articulation of thoughts, beliefs and wisdom, which the indigenous culture evolved and maintained as a part of day-to-day guidelines to their social activities, which were exercised with natural forces. It is through the experience with the environment human beings have articulated values through myths and have acclaimed strong social foundations in understanding the environment. Thus, oral tradition has formed strong theoretical edicts of Khasi ethics in relation with the environment.

The existence of deities and spirits in nature according to Khasi believe system is functional for good virtuous act of Man and can cause serious harm against any misuse and disrespectful act against the environment. It also relates with ecological sites and most of these sites like the 'Sohpet bneng', 'Lum Shillong', 'Lum Raitong', 'Law Kyntang' (Sacred groves at Mawphlang Dawki Cherrapunjee are the few known ones) Sacred River the Myntang Ka Kupli Ka Um Iaiam Ka Um Ngot are places which are the abode of spirits and deity which are accordingly good as well as evil. It is also the place where the 'Rngai' (shadow) of the deceased roam about and people should abstain from any disturbance to these spirits. In order to appease them, the Khasis perform religious rites and rituals with sacrifices. Based on such beliefs and practices creates an eco-spiritual tranquility between man and environment. It is a strong belief system among the Khasis who are collectively conscious about the environment. Their cultural practices have functional implication that is to hold and sustain Khasi solidarity. Their sense of collectivity is so strong that is almost impossible for a Khasi to breach away even in the whims of new modern theories of livelihood. Khasi philosophy, ethics and its customary practices are to be understood as 'social facts' as theoretically framed by Durkheim – Rules of Sociological Methods, 1895. Traditional practices are facts with moral rule and effective guide to control the conduct of man. Khasi morals are so deeply associated with ethics and philosophy executed in day to day activity which are external in form impresses upon its people the reasons can be traced in the Durkheimian perspective in which such execution of thought is not a simple imposition of outside control on an individual will but rather a moral obligation to obey a rule as Durkheim writes "Something beyond us and something in ourselves". This is a clear evidence in Khasi interplay between man, god, spirits and nature. Their relationship transcends beyond social purview with collective rules of thought as moral obligation.

Khasi socio-religious ethics is a complex embodiment in which composition of both flora and fauna as well as cosmic bodies are sacred. As an evaluation, the sacredness of environmental elements finds its space on Khasi, myths, fables and accordingly, they are personified with distinct qualities and power. Animals, birds and Man in the Khasi philosophical myths share a close commaradre expressed in Khasi literature writes "U briew U Mrad kawei ka ktien ki dang ia kren" (When man and animals spoke the same language). This happened in 'Sotti-Juk' the (Age of Purity) the primary stage in Khasi evolution.

Khasi rites and rituals use a number of natural elements like the water, rice, a rooster/cock, an egg, liquor, bamboo and water gourd. They are environmental elements which are used as symbols associated with ideas and meanings that connect man with nature, these elements are potentially live sustaining elements. It is the eco-consciousness of the Khasis to use them as a mark of sacred symbols, which evokes a deep spiritual sentiment in human perception about nature. They are used as physical objects of nature with its utilitarian value, but its religious meanings fulfil certain intrinsic attributes and meanings imparted to the object, which are used to fulfil those meanings and its religious implications.

Water (Ka Um) – Water is the Source of Life

Chemically composed of two atoms of hydrogen and one atom of oxygen its chemical formulae H_2O is an everlasting environmental and universal bliss to man and planet earth. Khasi livelihood have evolved with an experience encountering to it in every form of its sources – lakes, ponds, rivers, waterfalls, streams and brooks, spring and hot spring, cave water and rainwater.

Water is personified as 'Ka Um' as one of the child of Ka Ramew and Basa. It is not only used for domestic purposes but it is used as a fundamental source in any Khasi religious rites to connect man with the almighty. It is used for cleansing the altars and in many forms of divination. It is also used as a source of judgements in social life.

One of its customary rule practice in Khasi society (Sohra kingdom) known as Ka Jingingam Um (water diving) conducted in two ways. According to the first order, each contending party for seeking right judgement bring its own expert divers known as Ksih (seal) who is to remain immersed in water for a longer period of time to bring victory to his party. As per the second method two pots one containing gold and the other silver are placed at the river bed or lake. The divers dive into the bottom of the water, to fetch the pots, the objective of the diver is to get hold of the gold pot to bring favourable decision for his group. This water-diving ordeal for attaining right judgement is found in Jaintia hill as I Yangam um and I Yangam Ksih. Decision on such ground is considered final for water is assumed to be a sanctified as pure and powerful to have an authority over human beings. Here the decision is final and binding and if anyone is against the decision obtained on this account, it is equal to displease god and his choice. Social and customary as such are motivated on strong belief system of the Khasi.

Rice – U Khaw

Rice is another sacred and natural element and is the staple diet of the Khasis. Rice is an agricultural produce by dint of hard labour and honest means. It is used in each and every religious rituals of the Khasis in the form of rice grain, rice powder and rice paste. It is always considered to bring good fortune in the house and is also regarded to be a protective element if carried personally as a good charm.

The Rooster (The Cock) in Khasi Life

A Cock/Rooster in Khasi society is a mediator between God and Man. In many religious rites and divination, a Cock is sacrificed and its entrails are read by the diviners to find God's sign of agreement or disagreement to an event. A cock is used for religious rites with authority in the Khasi myths where a Cock is addressed as "U Khun Ka Blei uba kit ryndang ba shah ryndang na bynta jong nga u briew" (The son of God who gives himself up as a ransom for mankind).

The cock crowing symbolizes bright light of the sun in man's life, which fosters purity, humanity and divinity, thus, a cock composite of an environmental fauna makes a strong relativity between man and other living creatures, which are capable to streamline human action against Nature.

An Egg (Prayers and Divination with an Egg)

Using an egg in Khasi traditional rites is the sole in keeping Khasi customary practices intact. It is used in almost all the religious ceremonies. More precisely it is used in Khasi divination. Egg breaking divination entails a team-work of the diviner, an egg and the deing shat pylleng (an egg breaking wooden board) this board is not made from any ordinary wood or tree, it is only made from the log of a tree that have fallen on its own by a strike of thunder and lightning.

Before breaking the egg, the diviner places the red earthen soil on the board then little heap of rice on it and places the egg on the middle upright, and in deep concentration muttering incantations, invokes the God and deities and prays to the God Almighty who controls nature and man, he then sweeps away the rice and cleans the board thereafter holding the egg he prays to god for showing signs of god's agreement and guidance on the act as he dashes the egg on the board with considerable force. Breaking of an egg and it scattered shell which is upward is considered to be favourable and if it is downward it signify unfavourable for the event. The position of the shells is read by the diviner who prognosticates good times or bad times ahead. Using an egg is a strong belief of the Khasis that God witnesses the event and give accordingly his permission or non-approval through the execution of the egg.

Liquor – U ka'iad

Liquor is regularly used as a common drink among the Khasis and an essential requirement in socio-religious rites. It is processed out or fermented and distilled rice juice or millet call Ka'iad Pudka or rice beer which is of two kinds – Ka'iad Um and Ka'iad hiar. Beer made out of a root of a plant is known as Ka'iad Wiang. Ka'iad hiar is said to be stronger than Ka'iad Um.

Rice beer is used as offerings in rituals and when used for such purpose it is sanctified to be sacred.

Bamboo – U Siej

Bamboo in the life of most indigenous tribe is the cardinal environmental element used in day to day activities. It is unique in the study of socio-cultural evolution of men.

Bamboo is grass vegetation which belongs to the family of poaceae and its sub-family nomenclature is bambusoideae. Bamboo is used in many cultures of the world and has been marked in many research objectives to understand the process of eco-friendly activities.

Bamboo in Khasi society is a versatile plant, and it is interwoven deeply in Khasi life and culture along with bamboo canes a non-wood forest product and its genus calamus is used in crafts and basketry and in fact an important industry of the Khasi society. Baskets of different forms, fan, musical instruments, furniture, winnowing fan, rain shield fish and insect trap different type of mats, bows and arrows marks a skilled and creative art of the Khasis. These products are used in day-to-day life in Khasi livelihood. Bamboo tender shoots known as lung siej of several species of bamboos are of great demand used as vegetables and makes an excellent pickle and its juices for some medicinal uses. Bamboo planted out of 'Sharati' (flute) of Manik Raitong planted by him in an upside down position produces bamboos whose leave grows downward is noted for its fine quality.

One of the age-old practice of the Khasis in agriculture is known as "bamboo drip irrigation system", which is practiced by ancient Khasis with aqueducts called Shyngiar. By some Khasis this system is used for irrigation of betel leaves, Tympew Black pepper plantation and arecanut orchards called bri kwai, in the cultivation of these plants no metal based implements were used for the belief that Kwai and tympew are the sacred element of nature and require an extra care and hygiene for its harvest.

A Khasi uses bamboo just after birth, when the umbilical cord is cut by a sharpened bamboo splinter and no other tool of any elements were used. Thereafter, offering are made with an egg and water to certain an appease water deities 'ka blei sam-um' and 'Ka niangriang' – a spirit which is supposed to have the power of causing a disease of the navel of a child. They also pray to the forest spirit "U suid bri u suid khlaw". This is followed by the naming ceremony to seek blessing from nature.

Bamboo is also used in archery. Khasis are good archers. Archery originated with a Khasi mother "Ka mei ka nong hokum" (mother who commands authority), and her two sons. A man's life in Khasi world begins with an 'arrow ceremony' if the baby is a male child, a bow and three arrows (nam sop) are kept nearby indicating him to be brave as a warrior. Off the three arrows one is designated for self-protection, one for protection of his family and clan and the third one is to guard his territorial right. For the girl child a bamboo basket is also used in the 'Tang Jait' (ceremony for a new clan). In Khasi social system as per the religious rites, the girl child who will begin new clan is to carry a bamboo basket 'Ka khoh', the basket symbolically represents a composition of a Khasi tradition and matriliny as a form of lineage. It also symbolizes her to carry forward her new clan ethics and observe the traditional religious rites and duties. This is an obligatory task on her shoulder to fulfil throughout her life social and religious responsibilities.

During the funeral rites Khasis use bamboos which is called 'Japung' (bamboo mat) on which the corpse is laid for the ritual to be taken over and a small bamboo basket is hung over the head of the deceased. Three arrows are shots skywards one to the east, another to the west and the third to the north. It is believed that these arrows will guard the soul in the last towards god.

Thus bamboo is used to protect man since birth and provide all the essential means for a Khasi livelihood. It is symbolic to ascertain guidance, power, and authority and determines Khasi ethics and responsibilities.

The Calabash 'U Klong' – The Water Ground in Khasi Life

The Calabash commonly known as Water Ground belongs to the family of cucurbitaceae and its binomial name is Lagenaria siceraria.

In Khasi terminology, it is known as 'U Klong'. The calabash is a vine grown plant as a fruit/vegetable. It bears white flesh and smooth skin with seeds and sufficient water content as its juice. It is used as utensils, smoke pipe, water container, musical instruments in many societies all-over the world.

In the Khasi Society Water ground 'U Klong' is one of the cardinal elements in many religious practices without which religious ceremonies loses its essence.

The hollowed out dried Calabash 'U Klong' is sacred and highly used in many socio-religious rites.

Ceremonies Where the Water Gourd is Used

"Ka jer ka thoh" (Naming Ceremony).
"Ka Poi kha-poi man" (Marriage Ceremony).
"Ka Niam Iap" (Funeral Ceremony).
"Kaba tep Mawbah bat tei Mawbynna Niam" (Interment of bones and stones erection ceremony).
"Ka Tang Jait" (Creation of a New Clan Ceremony).

In all these ceremonies U Klong is used to contain water but more often liquor out of fermented rice juice, which is to be sprinkled or dropped down the liquor teaming up the verse cheated out by the elderly local priest (*lyngdoh*). This part of every religious rite is a sacred ordeal which are observed and performed in the presence of the members of the traditional institution and the villagers. If a man has to synchronize in full concentration and pray holding the gourd as an object of pure and sacred, this brings a deep attachment to it on spiritual context making requests for the blessing from the Almigty god "U Blei nong buh nong thaw" (god the originator and creator). The significance of this act is the divine connection between Man, Nature and God, such practices makes a humble water gourd so sacred that the ritual should be solemnized with utmost seriousness and care and any follies in the event is believed to cause harm to both Man and Nature.

Likewise there are many other natural elements which are used to procure their utilitarian aspects and are defined in a sacred note, this makes the society use nature and its resources with fear and respect that paves a reason for protecting environment. It is a strong belief system of the Khasis the relation between man and environment is rooted in religious context, and this is held collectively to achieve the objective utilitarian function of natural resources on the basis of sacred note. Traditional practices have been constant among its new generations for it is indispensable in Khasi society. The environment is so intensely entwined in Khasi philosophy and social habits where both have synthesized into forming people's identity and its cultural ethos. In the words of H.O. Mawrie (translated by R.M. Shangpliang) "rituals are a part and parcel of our culture. They have embedded their roots deep down in our flesh and blood and to think of mankind to throw away either of them is to tear oneself apart".

Forest and the Khasi Society

Forest is the gift of god to the Khasi society. Khasi culture – its evolution, civilization and development teams up with forest. Its broad spectrum includes food, shelter, medicines, agricultural implements and far beyond these it is educative and gives a sense of ecological and economic security.

The term 'forest' has been derived from the Latin word 'foris' meaning 'out of door'. Etymologically, it is "a large uncultivated area of land covered with trees and underwood" (*Chamber's 20th Century Dictionary*, 1943). According to the *American Heritage Dictionary of English Language*, a forest is defined as a "dense growth of trees, plants and underbush covering a large area". The relationship of man and forests is an age-old history in human civilization.

Forests in Meghalaya are broadly grouped and classified into different types.

1. **Tropical Forest:** These forests are within areas up to an elevation 1200 m and with an average rainfall of about 100–200 cm. They include numerous sub-types such as – (a) evergreen in this category Tura range up to Siju forests is filled with nature's extravaganza species here are mighty and the forest is covered by tropical flora; (b) semi-evergreen: Tropical semi-evergreen forests this type covers the North-eastern slopes of the state. It also covers the areas of Saipung and Narpuh forest. Trees found here are Careya Arborea, Dillenia, Pentagyna and Callicarpa Arborea; (c) Tropical moist and dry deciduous forest: This type of forest occurs with annual rainfall below 150 cm. It is characterized by seasonal leave shedding and flowering trees. Trees like Shorea Robusta, Tectonagrandis etc. Bamboo is grown abanduntly.
2. **Grass Land and Savana:** This type of forest covers Shillong Plateau Riangdo, Ranikor, Weiloi, Mawplang, Mawsynram, Cherrapunji, Shillong-Jowai, Jairain and Sutnga. This type of forests support other Dicotyledonous Species.
3. **Sub-Tropical Pine Forest:** These forest cover Shillong plateau and Pinus Kesiya is the tree grown in this forest.
4. **Sal Forest:** These forests occupy northern part of Khasi Hill and some parts of Garo hills. The tree grown here are Shorea Robusta known as *sal* tree.

Forests and Its Functional Uses in Khasi Society

Khasi social livelihood rotates round the forests and its everlasting bliss on man. Khasi folklore and legends are created with a close symbiotic relationship with the environment. Khasi legend of 'U Diengiei' a giant tree evokes strong moral about good deeds and its rewards always superseding bad deeds and its bad rewards. The legend of 'Sier Lapalang' depicts the beauty of forests as well as the difficulty and disaster one faces of being disobedient and crossing limitations of your freedom. Such fables and folklores convey the importance of values in life and their spiritual strength of the story passes a message to humbly accept simplicity of life close to nature.

Forest resources are also used for making houses. Traditional houses are made of grass, bamboos, planks and mud are forest's resources. Houses of the Pnar-Wars uses leaves of a palm called 'U Tynriew'. The Bhois Lyngam houses are made of bamboos which are 30-40 feet in length. In the Khasi society use of nails in the construction of houses is tabooed 'Sang'.

Trees Used for Making Thatched Roof

- *Caryota urens* (Dieng Tlai).
- *Cymbopogon exertus* (Langphriet).
- *Oryza sativa* (Skum).
- *Saccharum spontaneum* (Kdiat Miaw).
- *Themeda caudata* (Sder, traw tdie).

Trees Used for Making Walls

- *C. traninhesis.*
- *Leea guineensis* (Dieng Lang Karnum Khythei).
- *Thysanolaena maxima* (Lynnai).

In Bhoi and War region bark of 'Kalatar' as high as 70-80 feet is used in tying poles for constructing houses. Pegs made out of bamboos are also used instead of nail, the example of such a house still exists in the Ing Sad at Smit.

In the world of aesthetic Khasi cane and bamboos product are used in making variety of basket and furniture as well as other agricultural and household essentialities.

Trees Used for Making Furniture

- *Albizia lebbeck* (Deing Soh Rin).
- *Bursera serrata* (Deing Soh Mir).
- *Carallia brachiata* (Deing Soh Langbali, Soh Syllih.
- *Gmelina arborea* (Deing Laphing, Lophiang).
- *Lagerstromia speciosa* (Deing Ther).

Bamboos and Canes Used for Making Baskets and Furniture

- *Nutans* (Serinjai, Seijranjai).
- *Tulda* (Seij Rangtah, Seij Skong).
- *Dendrocalamus Gigantus* (U Ktang).
- *Neomicrocalamus prainii* (Sampit, U Speit).
- *Collinus* (Thri Phin).
- *Leptospadix* (Thri Saw).

Bamboos and cane are one of the recourse for cottage industries and are in high demands.

Forests produce have motivated and inspired the Khasis in producing musical instruments for which hard wood, animal skin and bamboos are used.

Trees Used for Making Musical Instruments

- *Aralia montana* (Deing Latymphew) used for making traditional *Dvitara.*
- *Artocarpus heterophyllus* (Deing Sa chram) for naking traditional Drum.
- *Sacchrum benghalense* (kdiat).
- *Schefflera hypoleuca* (Deing Lakhor).
- *Viburnum foetidum* (Deing Soh Làng, Soh Lang Skiew) for making Tangmuri.

The art of weaving and dyeing as a part of Khasi culture highlights eco-consciousness of the Khasis towards aesthetic sensitivity. Colours are extracted from leaves, roots, fruits, stems and barks of trees from the forests.

Plants and Trees Used for Herbal Dyeing

- *Albizia odoratissima* (Deing Krait Iong, Deing Saw) Bark and leaves are used.
- *Aporausa aurea* (Deing Soh Kyrsiew) Leaves are used.
- *Bischofia javanica* (Deing Soh Tung) Leaves are used to dye bamboo baskets.
- *Curcuma longa* (Synrei syrmit) used for dyeing silk.
- *Datura stramonium* (Soh Siah) Fruits are used.
- *Girardinia diversifolia* (Thiang Thap) Seeds are used.

Trees Used for Weapons and Other Essentialities

- *Aquilaria khasiana* (Deing Akhar).
- *C. tamala* (Deing La Tyrpad) handles of wait.
- *Castanopis indica* (Deing Soh Ot) handles of axes and mortar.
- *Aracia sinuata* (Jyrmiersa) for washing purposes.
- *Byttneria grandifolia* (Jermi, Mei Klong) used for washing hair.
- *Elsholtzia flava* (Ja Thang Niang) used to kill bugs.
- *B. nutans* (Seij Rynjai) for water pipes.

The Khasis have knowledge about forest plant, shrubs and many other resources for curing many sicknesses. Traditional knowledge about herbs and its medicinal value of the Khasis have been passed down the line from generation and generation, which they called as 'Kynbat' and it is effective in curing some of the ailments, as bone fracture, food poisoning, tooth ache, burn etc. Today with the development of science and technology few Khasis have opened up their own health centre and have cured chronic ailment, these are scattered all-over the villages of Smit and Nongstoin for example Dr. John Herbal Clinic in Nongthymmai and his hospital in Thangsnging Smit, Dr. john has patients and centre abroad as well.

Plants, Trees and Herbs Used in Making Medicines

— *Abrus precatorius* (Soh Rati) for curing jaundice.
— *Achyranthes aspera* (Soh Byrthied) for antiseptic.
— *Asatibum* (Rynsun) for hypertension and influenza.
— *Alpina calcarata* (Latara) for tooth ache.
— *Amomum aromaticum* (Kynshiang) for nausea.
— *Andrographis paniculata* (Dawai Wieh) for liver problem.
— *Aristologhia cathcartii* (Kynbat daitnud) for heart burn.
— *Asaccata* (Patisang) for spleen, stomach, and urinary trouble.
— *Atagala* (Soh Rynkhiah) for bone structure.
— *Artemisia nilagirica* (Jarilut) for malaria.

Forest for the Khasis is the 'food basket'. Foods as roots, tubers, mushroom, bamboo shoots, creepers etc. are yield by forests produce, forests also attracts insects for which api-culture is an age-old tradition of the Khasis. Variety of honey are processed and Khasi honey is marked for it good quality.

Conclusion

Today in fast developing, globalizing world and modern set-up, Khasi social lifestyle is neutralizing to the changing demands. The growing Western culture and its impact on the new generation have possibly created a gap between the indigenous traditional knowledge and modern way of life. Sophisticated technology, and newly defined gadgets have not only positively benefited the society, but the bottomline in its extreme usage have made our purpose dependent and less innovative and inventive. Tribal diligent behavioural attributes is been swept away from its culture owing to the greater attraction brought by the globalized consumerism lifestyle. Modern globalized action and its trends have engulfed North-Eastern way of life so deeply, which poses a great challenge for traditional knowledge, art and skill, which once evenly synthesized meritoriously with nature.

The growing challenge further lies in the fact that there is a lack of adequate knowledge about the indigenous properties in the environment and their benefits by the new generation, the problem to understand this matter is rather historical, to take note since the time with the advent of the British colonial hegemony. English way of life was then seen as the mode of standardized behavioural cult. Christian missionary educational institutes introduced English as the medium of instruction which was then viewed as positive opportunity for positive change. However, it did bring some benefit to the people in drawing out of some of the superstitious conceptions about natural and social phenomena which are based on Science and logical explanations. About 60-70 per cent of the Khasi population today does not subscribe their theories on unscientific and irrational explanation on any matter or on any phenomenon. Hence, traditional wisdom on environment and its theories are amidst greater challenge to prove with authentic interpretations based on logically right and true to its account.

Moreover, English as widely accepted language to be taught in schools at elementary level in most of the educational institutes are actually a threat to learning in local dialect. Traditional social and environmental knowledge are often transmitted orally from one generation to another in local dialect, very often many important meaning of the environmental and social terms along with their related implications are usually intact in local terms and language. The fear lies when language starts degenerating, the loss of indigenous traditional knowledge begins to ward off with the inability to use the mother tongue effectively and this have been a serious problem in North-East indigenous environmental and social knowledge.

Aspiring jobs, career are yet another issue of tribal culture widening the gap between the past and the present. Moreover, inadequate economic system and weak employment generating infrastructure in the region have forced migration of youths from rural to urban areas and from town and cities to other regions in search of jobs, which is again cutting away the roots from the tribal source, moreover being for long out of one's culture one loses the interest in local language and local practices. To regenerate the motivations is also rather practically a setback in striving for progress and modernity and at the same time maintaining once indigenous cult of thought and practices, most North-East youth at some point or the other are facing this issue.

Thus, the new budding generation and the societies at large in North-East India are in a state of confusion for integrating tradition and modernity in which both are dynamic and competent in nature.

What is to observe here is, without tradition one loses the root and identity and without modernism one also loses progress and growth, such a situation is inhibitive to decide and come to a conclusion. For instance modern medical science states that to chew tobacco and betel nut causes serious oral infection while on the other side chewing areca nut produces juices in mouth which are actually enzymes that helps in digestion of food for which traditional eating areca nut and betel leaf was encouraged to eat after food, which highlighted traditional knowledge and wisdom of the past in which predecessors gained such insight only by eating and experiencing the dynamic chemical composition of areca nut. Such experiences on many issues were further articulated ethical sources in establishing customary dictum. Such instances are many in indigenous knowledge which needs to be scientifically interpreted. Today's global economy lays much emphasis on human productivity rather on human knowledge of the past in relation with nature in harmony. Human productivity emphasizes globalized philosophy of survival strategy on the bases of progress, career, documentation of achievements, diplomatic qualities, objective approach, specialization and skills – all these qualities have no doubt upgraded materially, but is also degrading close contact with natural environment and its underlying meanings. Today, material achievement is the hallmark of every society which is also a part of North-Eastern socio-economic system and this invariably is constraining for the growth and maintenance of the past knowledge.

However, realizing the utmost need and importance of nature, educated and intelligent elites of the society have organized initiative forums and strategies to conserve, promote and protect indigenous tribal culture and wisdom, as an observation such featured exhibition and demonstration are encouraging and good enough to revive old traditions yet the fact is to respect Tradition one has to live in 'Tradition', be in 'Tradition' and practice 'Tradition', which modern set-up is practically not feasible in nature, yet the source of Indigenous practices is potential enough to forgo arising challenges, in which change with adjustment and adaption with integrative method helps in retaining traditional knowledge paving ways for new theories. Until the complexities of deep spiritual, social and environmental essence are explained by the people in the society, with a deep sense of awe as a moral guiding force traditional culture will continue to face era of any crises and challenges.

REFERENCES

Ahmed, Ashraf Ayesha and Borthakur, S.K. (2005). *Ethnobotanical Wisdom of Khasis (Hynniew Treps) of Meghalaya*, Dehra Dun: Bishen Singh Mahendra Pal Singh.

Aron, Raymond (1967). *Main Currents in Sociological Thought II*, Garden City, New York: Doublday & Co.

Clifford, Geertz (1973). *The Interpretations of Cultures, Selected Essays*, New York: Basic Books Inc. Publishers.

Gordon, P.R.T. (1975). *The Khasis*, Delhi: Cosmo Publications.

Mawrie, L. Barnes (2009). *The Khasis and their Natural Environment*, Shillong: Vendrame Institute Publications.

Nongkynrih, Sing Kynpham (2007). *Around the Hearth*, India: Penguin Books.

Shanpliang, M. Rekha (2010). *Forest in the Life of the Khasis*, New Delhi: Concept Publishing Company Pvt. Ltd.

Shanpliang, M. Rekha (2013). *Socio-Economic Dimensions of Forest Among the Khasis*, New Delhi: Concept Publishing Company Pvt. Ltd.

Other Sources, some footage of live coverage of cultural event at Nadia's Communication Library.

15

Man-Nature Relationship of *Borok* Community in Tripura

Jawahar Debbarma

Introduction

The relationship between humans and the environment requires some ideas of myths, traditions, religion, cultures and philosophical, political and economic systems in order for it to be explicit. This relationship is a vast area of research for ecology, particularly its cultural and ideological dimensions. The human-nature relationship also involves elements of philosophy as it concerns human kind's place in the world. Thus, those communities, particularly tribes, whose whole existence evolves in and through nature are definitive of this relationship. In short, their existence defines the relationship.

In this chapter, I have tried to look at the Tripuri (*Borok*) tradition, culture, their belief system and their relationship with the environment or nature. The web between tradition, culture and environment is implicitly interwoven in the lifestyle of people of Tripura. The understanding of Tripuri culture and tradition is incomplete without a brief description of the *huk* (shifting cultivation) practiced in the hilly areas of Tripura, for the entire traditional socio-cultural life of the *Boroks* is mirrored in the *huk*.[1] The *Borok* community and their life cycle were totally dependent upon *huk* cultivation. Birth, death, social rituals, music, dance, weaving, handicrafts, ornaments whatever the requirement of human life was derived from *huk* and forest resources. Thus, traditionally, I would say they are very close to nature and live in nature as their socio-economic and cultural life is based on the *huk* (shifting cultivation). For instance, where there were no clock or watch in their society, we find that in their traditional time measuring method certain natural signs were used to lead their daily life by which their close association with nature may be indicated and proved easily. Right from early morning time to the dawn time several natural signs are used in their everyday life in order to maintain time round clock. When *tokprengsa* (a kind of bird) cries it indicates that it is early morning, when cock crows it is understood that it is morning time, when cock crows very loudly again and again it tells that the sun is rising, when *khumbar* (a kind of flower) blossoms it is early noon time, when the shadow of a person falls on his or her feet straight it is noon time, when *tokpupu* (a kind of red brown bird) cries it is afternoon time (about 30 to 40 minutes before the sunset), when *jinga* (a kind of vegetable) flower blossoms or there is evaporation of water drop inside the bamboo tender leaves or *monai* (a kind of bird) cries, it is sun-set, when *sanjadoroknaima* (one evening start called Venus) is seen in the sky from the east direction it is late evening, when *swkwithakdari/takhukthamathukuri* (3 stars brother) are seen in the sky, it is *sanjakuthuk* (night), when *palabadukhung* or *takhuksiniathukhuri* (7 stars brothers) are seen in the sky it is deep night time, when *aitorma* (the morning star) is seen in the sky in the west direction it is early dawn time etc.[2]

According to scholars, the tribals or tribes of Tripura were divided into different groups based on their respective lifestyles. It was the king of Tripura who had led them to categorize themselves into different sections according to their lifestyles and this categorization was later recognized as a clan. It is generally believed that these clans are named after particular leaders, heroic deeds, professions, banks of rivers and places or village to which the group originally belonged and reseeded. The '*Dhopha*' or Clan of Tripura Tribe is further divided into twenty-one groups. These Clans came from the twelve Tripura families and they were known from the twelve sons born into the house of Trilocan. As cited in the *Rajmala*, Trilocan was born of Hirabati, wife of Tripura after his death. Tripura is the son of *Daitya*, the descendant of *Drahya*.[3] Each of this Clan is sub-divided into different '*Sandai*'. The *Kokborok* word '*Sandai*' literally means lineage. It is believed that the *Sandai* is named after some animals, birds, bamboos, waterfalls etc. as most of the groups and sub-groups of the *Kokborok* speaking tribals of Tripura had. But with the arrival of modern time, Tripuris have lost the meaning and characteristics of their *Sandai*. However, this fact suggests that these tribes had great respect for the respective group of birds and animals. In this respect resulted in less damage to the environment as well as helped in the preservation of the natural bodies. It is found that Tripura's lineages are patriarchal. The members of the different lineages are known, according to their respective ancestors.

Coming to the specific rituals of the *Borok* people, I discuss their basic rituals and traditions which exhibit the man-nature relationship as interdependent. First of all, are the rituals of birth. When a baby is born in a *Borok* society certain ritual is being practiced known as purification of the baby. The newborn baby is purified twice after birth. First purification is done on the day of birth, which is known as '*Abur Kisi Sumani*'. This ritual of worship is done by the *Ochai* (priest) by the river. The *Ochai* prepares a *binda* by sand, in the middle of the *binda* he puts *mairongchopaha*, a pair of pat (it is of two leaves and the leaves for pat should be barren of flowers and fruits), a piece of *wa-chook* (head of bamboo) tied with *khum-twrang* (flower garland) and a cock as per new born baby male or female respectively. It is believed that by looking at the intestines of the cock, the *Ochai* can predict about the good or bad destiny of the new born baby. It is said, right hand side is for god and left hand side is for man. So, it is decided accordingly that if the *Ochai* found the intestine of the cock in the left side is long, it means good for the new born baby, otherwise it is not in favour of god and this symbolizes evil. The second purification is also worshipped and practiced same as *Abur Kisi Sumani*, but it is called as '*Abur Kwran Sumani*'. On the seventh day after the child's birth, name giving ceremony takes place. The name of the child is usually given relating to the nature like *Khumbar* (*khum* – flower, *bar* – bloom) *kherenbar* etc. These rituals of birth depend hugely on the surroundings and the nature. River is essential, besides leaves, flowers and fruits of certain types. The cock also has a significant place in this ritual. The connectedness to nature is not a mere ritual; rather, it is what shapes the very existence of an individual. Thus, any disrespect to nature, trees, flowers, animals etc. is seen as the violation of the spiritual and environmental values that hold the *Borok* people together.[4]

Secondly, when death occurs in a Tripuri family certain rites and rituals are observed. Although, these rituals and rites are highly influenced by the Hindu rituals and ceremonies, nevertheless they have become the very part of tribal in Tripura. They have completely overshadowed the early tribal rituals, which we can no longer trace in current practices. Yet, these rituals are not far from the environment. First the relatives of the deceased person, family members are informed then followed by the preparation of funeral bier, the bier (*Talai*) for the dead body is made of bamboo. The dead body is kept on the newly made bamboo platform. The head of the dead body is directed towards North. The relatives and other persons who come to pay homage to the dead body give a small amount of paddy, cotton, sesame etc. at the feet of the dead body. After that, a cock is killed at the feet of the dead body by striking in on the ground with the left hand, thereafter; rice is cooked in a bamboo pipe. The cooked rice and that chicken are offered in a newly made bamboo basket for the dead body. In that particular basket four symbolic doves are made with bamboo. Two doves are in eating positions and the other two are positioned to fly away. Now, the funeral procession proceeds towards the burning place carried by his sons, son-in-laws and the family friends then the dead body is cremated. As soon as the dead body is cremated, the participants in the cremation of the dead body return back to the deceased's house. On their way back home, they tie a knot in the tree, leaf or in plants by stopping the breath. Everybody is strictly prohibited to look back. Small hut '*austi nok*' is also made nearby the mourner's house symbolizing the hut for the dead body. The mourning period is observed in the deceased family for twelve days and during this period, no socio-religious activities can be performed at the house of the mourners. During the first three days of the mourning period, the offering of the foregoing system along with a new cloth is offered to the departed soul before the sun rises. This same cloth is offered repeatedly for consecutive three days. On the fourth day, the son goes to the place where the dead body is being cremated and collects the skull of the dead and later immersed in the river. The duration of the mourning period last for twelve days and on the thirteenth day they perform the '*Maikhalai*' (*Sraddha*) ceremony – the offering of food and drink to the departed soul. During this mourning period, the family members and the relatives of the deceased are not allowed to take fish, meat, onion, boiled rice etc., only for twelve days after the *Maikhalai* ceremony on the thirteenth day the family members and relatives break the norms of taking vegetarian food and take the fish, meat etc., or non-vegetarian food, all this ritual is performed with the help of the *Ochai* (priest). Death rituals usually end with the rites of *Khumbokwi* of the middle month of March and the beginning of April. Though the ceremony performed during death has Hindu influence, it is basically traditional. The belief system accompanying, it reveals the importance of environment. The use of bamboo is an integral part of this ritual. Besides this, the sense of living even while dead, suggests the hesitation to destroy river, mountains and forests. For these places fill a tribal with a sense of fear and awe.

To draw our attention on the religious practices and aspects of the religious belief system, and thus to get a connection with nature, I now elaborate on deities that guide peoples' relationship with the nature. The *Borok* tribes have a good number of popular god and goddesses. It is worth mentioning here that among the Tripuris there is no specific form or shape of the deities that is found. Thus, unlike Hindu gods and goddesses, deities of *Borok* people are different and should not be equated with Hindu gods as Prof. Jena does in his book.[5] In most of the cases, the structure of the deities is made with bamboo poles or stone. The altar of the deity is made with two bamboo poles which are erected on the ground. Along with these two planted bamboo poles, two other bamboos are tied to each other horizontally for specific deities. To make the structure of the deities only the specific green bamboo called '*Muli*' variety is needed. Sometimes even the stone is also worshipped.[6] Among the deities of Tripuris, some are benevolent and some are malevolent which have a great influence in their life. The malevolent deities or spirits are held to wreak havoc to both mental and physical life. And therefore, Tripuris are very careful to appease the hostility by sacrificing animals and birds. Each god and goddesses are addressed as '*Ma*' and '*Pha*' which means 'Mother' and 'Father' respectively.[7] It is good to remember that the sea, the sky and the mountain seem to be the gods and the earth, the water and the tree become the goddess. The Tripuris worship them in different places with different names by the help of *Ochai* (priest) or without *Ochai*. The offerings to the gods and goddesses include *Dali-Kamchi, Chuak-Kathar* (holy wine) and domestic animals like goats, chickens, pigs etc. Some of these benevolent deities are as follows:

1. *Matai Gorja* (*Matai* = God; *Gorja* = Chief), *Matai Gorja* is the supreme god believed by the Tripuris. It should not be misunderstood with *Lord Siva*.
2. *Twima/Twibukma* (*Twi* = Water; *Ma* = Goddess), *Twima* is the goddess of water.
3. *Mainungma* (*Mai* = Paddy; *Ma* = Goddess), *Mainungma* is a goddess of fortune, wealth and prosperity. This is not *Lakshmi*.
4. *Khulnungma* (*Khul* = Cotton; *Ma* = goddess), it is a goddess of cotton.
5. *Hachukma* (*Hachuk* = Mountain; *Ma* = Goddess). It is a goddess of hills and mountain. This in not *Parbati*.
6. *Noksu Mwtai:* It is a household deity.
7. *Kalia bai Goria:* God of agriculture and crops.

8. *Lampra Uathop:* God of sky, path and sea. It is worshipped for the well-being of the self as well as society.[8]

The *Borok* people had built their philosophy in close association with their environment, where the man-nature relation is a sacred one. Nature is a priority, and man is an essential part of nature. The man-nature relation can be discussed and pointed out through studies of their myths, legends, folk tales, folk songs and even their superstitious beliefs. Even in the recent times, it is observed that a good number of Tripuris living in the interior villages are having superstitious outlook and therefore, we look at the few instances that depict their philosophy.

I. In the forest where the deer eat soil the Tripuris never cultivate '*jhum*' there, since they believe that if anybody cultivates '*jhum*' in that place any untoward incidents may take place in the family. According to them any deer having received any injury in its body if somehow can reach that place, then it is able to heal its injury by eating the soil. This shows the respect for the injured animals in par with injured human being.
II. The Tripuris believe that whenever a hen crows like a cock any type of danger may take place within the family in the near or distant future.
III. On the first day of paddy plantation many Tripuris till this day maintain some superstitious outlook. On that day, they do not eat '*Chakui*' (a kind of curry prepared by the filtered water from ash) and any sour thing. Soap is also not used on that day. The aforesaid prohibition is observed on the basis of the belief that the paddy will be less productive if they do not follow.
IV. If any passerby sees a snake on the right side of the path, then it is taken as a bad symptom.
V. If anybody in a family dreams elephant approaching the house, then it is believed that paddy cultivation will be much better that year.
VI. To dream to extort a cow or buffalo is believed to be a good omen for the family.[9]

Of course in this modern age these beliefs cannot be taken seriously, but nevertheless, they signify a specific role that these belief systems play in conservation of nature. They suggest that people created their worldview close to nature, deeply connected with nature and instilling in their children a sense of awe and wonder at nature. This sense of respect for nature resulted in restrictions on, unmindful usage of natural resources. These superstitious beliefs indirectly led to the preservation of forests, trees, mountains, rivers, animals, birds and nature as a whole.

The folk songs of the Tripuris also reflect an interaction with the nature. The Tripuri folk songs are as old as the tribe and they have survived through the ages as a tradition. The folk songs of the Tripuris were composed in the early days of their collective lives of individuals whose identity are unrecognizable and ignored. These folk songs are based on the old tradition, thoughts, desire, love, *jhum* cultivation, harvesting, festivals, beliefs and superstitions etc. from the folk songs; one can have a glimpse of the ancient constitution of the tribal society, its environment and livelihood. Some of these songs are *Jadukalija*, this song is based on love affairs, *Resekhagra*, this song is usually sung with a plaintive tune for commemorating a person after death. *Owaing Khilimani* is a song of lullaby usually sung at the time of lulling a child. *Mamita Rumani*, this song is sung after keeping the new crops in the granary, a ceremony is celebrated which is called '*Mamita*'. An example of such song is the following where the mind finds joy and solace to see the beauty of the nature:

> "*Bolong kwchangwi tonkha*
> *Hachukgo huluksa pungkha*
> *Nokbar bai bwlai narjakma nukgwi*
> *Khapangbo narjak paikha.*"[10]

It means the forest is looking so peaceful and beautiful in all dimensions. On the hill the owl is hooting. The tree leaves are shivering in the air. My mind is also dancing with joy at this sight.

> "*Ani hukchunggo sabo huk hoknai*
> *Wathwi kwlaio rai rai*
> *Khani borokno sabo kok rwfai*
> *Muktwi kwlaio roi roi*
> *Hopung hor thuwi muktworwoi maiya*
> *Owansukmang bwkha paia. . . .*"[11]

Here through this song the boy expresses his mental agony related to the hurdles in his love affairs. It pains to know that somebody else has negotiated with his beloved for marriage and his heart aches. And in his agony, he sings the song where he compares his beloved with the wild flower which so high to get hold of.

"Kuchuk kherengbar buabar motomno
Bahaikhe khollangnani
Bolong boltuku nana faijakgoi
Jungjalo kwlaimani". . . .[12]

(How shall I pluck the *Kherengbar* flower, which has bloomed so high? After coming to the forest for collection of firewood, what a danger I have fallen). It is a love song and it signifies that both the boy and the girl love each other, but they cannot come closer due to some inhibitions. Here the boy expresses his fascination for the girl and frustration of his love.

"The *Borok* people extended the idea of respect to the natural world not because they conceived it as an extension of the human community. The natural world of the *Borok* is, of course, inspirited. The legends, myths and folk tales, folk songs describe animals, trees, mountains, rivers, clouds, sky as beings who felt, perceived, deliberated and responded voluntarily as persons. This however must not lead to the conclusion that they saw the natural world as populated with other persons, persons like oneself. On the contrary, there is a great difference between animals, trees, mountains, rivers and us. But the difference does not amount to their being devoid of consciousness and intentionality. Each is a different species, though a member of the same biotic community one must respect the members severally".[13]

Conclusion

With the advent of modern era tribals are forced to come out of their isolated lifestyle and interact with the world of others. They are in contact with many other tribals and non-tribal communities. Due to the closeness of the tribals with the other neighbouring communities, distinct changes are seen in their traditional way of life in all respects. Their present style of living is significantly deviated from their traditional and customary ways and represents an admixture of tradition and modernity at present. Some of the major factors for these changes are economic, environment, globalization and urbanization. The economic condition of the *Borok* community can be considered as one of the most important factors which brought many changes in their traditional pattern of lifestyle. It is observed that the economic pattern has undergone many transformations in *Borok* community. As in the past, *huk* cultivation provided a self-sufficient economy among the *huk* cultivator's family. There was enough land for *huk* cultivation as the man-land ratio is equally balanced. Daily market was not found nearby their villages. Essential commodities like rice, fish, vegetable, cotton, jute, firewood etc. are not required by them to purchase from the market. Their basic items to purchase from the market are *berma* (dry-fish), salt, oil and kerosene. However, this situation has changed after Tripura joined the Indian Union on 15th October, 1949. The merger has brought about thousands of refugees from East Pakistan (now Bangladesh) and again thousands came over during the Bangladesh Liberation war during 1970s and were given rehabilitation in Tripura. The impact of such illegal migrants has changed the entire demographic situation of the state. They have occupied a major portion of the lands for the *huk* cultivation. The refugees also have outnumbered the Tripuris in some areas. The ratio of the land and people also showed a distinct change. Now, there are too many people and too little land to accommodate, thus, leaving less land for the *huk* cultivation. Since then the cycle of *huk* cultivation could not be carried out as before and the cultivation suffered to such an extent that it is beyond recovery. This has affected badly to the *jhum* cultivators and their self-sufficient economy and gradually shrunken their source of income, leading to traditional and cultural decadence.

Table 15.1: Trend of Population in Tripura (1874-75 to 2001)

Year	*Total Population*	*Non-Tribal Population*	*Tribal Population*	*% of the Tribal*	*% of the Non-Tribal*
1874-75	74,523	27,000	47,523	64.00	36.00
1881	95,637	45,722	49,915	55.00	45.00
1891	1,37,442	67,150	70,292	51.00	48.00
1901	1,73,325	81,646	91,679	52.89	47.00
1911	2,29,613	1,19,484	1,10,129	47.96	52.04
1921	3,04,437	1,37,937	1,66,500	54.69	45.31
1931	3,82,450	1,79,123	2,03,327	53.16	46.84
1941	5,13,010	2,56,019	2,56,991	50.09	49.91
1951	6,39,029	4,01,071	2,37,958	37.23	62.77
1961	11,42,005	7,81,935	3,60,070	31.53	68.47
1971	15,56,342	11,05,798	4,50,544	28.95	71.05
1981	20,53,058	14,69,288	5,83,770	28.44	71.56
1991	27,57,205	19,03,860	8,53,345	30.95	69.05
2001	31,91,168	21,97,742	9,93,426	31.13	68.86

Sources: K.N. Jena *et al.* (2009). *Life and Customary Laws of Tripura Tribe*, pp. 11-12.

The environmental factors have also brought some changes as in the past the Tripuris villages were usually surrounded by mountains, rivers and forests. Various kinds of vegetables are produced from their *huk* cultivation. They hunted wild animal, collect firewood, bamboo shoot etc. from the forest. The forest was very close to them as it provides all their food items to sustain their life. But due to globalization, urbanization, deforestation and overpopulation have brought an ecological unbalance towards their traditional way of life. The forest, which is precious and connected to their lives, no longer exists due to deforestation. As it is felt they were thrown from the lap of the nature. They now have to purchase bamboo, thatch, tree etc. for the construction of their houses, which were unthinkable to them even forty years ago. The extinction of the vast forest gradually led to the extinction of the *huk* cultivation due to which the Tripuris folk song, folk literature, folk administrative and folk festivals which has developed and flourished around the forest and *huk* cultivation has gradually declined. Dr. Ganguly rightly said "primitive people living in isolation, develop their own customs and socio-economic institutions which are the results of hundreds of years of gradual adaptation to the environment. So, any abrupt or unplanned change in their environment disturbs that equilibrium and causes maladjustment to the discomfort of tribal folk. This may even lead to their degeneration".[14] The other danger lurking to the tribal life and tradition is rampant rubber plantations. Although, it is aimed at capitalizing the area and increase of family income by the government initiative in the 1960s, it has caused havoc to the tribal mentality and their environmental philosophy. Profit has become the main objective behind this agricultural venture. And since profit becomes the centre of human endeavours, environment and cultural traditions are set aside or tossed off, resulting in a degraded social, cultural and environmental way of life.

To conclude, I would like to say that modernization and globalization has taken the world away from the environment. Tripuri tribes too, have adjusted themselves with the Hindu as well as Christian customary practices which differ from their traditional customs, rituals and rules of the abodes. But the traditions of the *Borok* community seem to evolve their own philosophy for contemplating the sacred harmony between man and nature. Mohan Debbarma writes "that their environmental philosophy is nature-centric, but not anthropocentric". He continues, "animals and inanimate things are also personified in the indigenous tribal community, they are all interdependent. Their traditions are inseparable based on respect for religious and cultural veneration of the natural world".[15]

NOTES

1. Debbarma, Mohan (2014). "Environmental Philosophy in the *Borok* Community of Tripura", *Nature, Culture and Philosophy*, Saji Vargese (ed), Delhi: Lakshi Publication, p. 222.
2. *Ibid.*, p. 224.
3. K.N. Jena *et al.* (2009). *Life and Customary Laws of Tripura Tribe*, New Delhi: Abhijeet Publications, pp. 29-30.
4. Debbarma, Mohan, *op. cit.*, p. 222.
5. Jena, K.N. *et al.* (2009). *Changing Face of Tripura Tribe*, New Delhi: Abhijeet Publications, p. 86.
6. Debbarma, Debapriya (1983). *Treatise on the Traditional Social Institutions of the Tripura Community*, Agartala: Director of Welfare for Scheduled Tribes and Scheduled Castes, p. 55.
7. Jena, K.N *et al.*, *op. cit.*, p. 86.
8. *Ibid.*, pp. 86-87.
9. Debbarma, Debapriya, *op. cit.*, pp. 58-60.
10. *Ibid.*, p. 77.
11. *Ibid.*, p. 79.
12. *Ibid.*, pp. 79-80.
13. Mohan Debbarma, *op. cit.*, p. 227.
14. Bhattacharjee, Suchintya (1992). *From Jhuming to Tapping*, Agartala: Director of Welfare for Scheduled Tribes and Scheduled Castes, p. 60.
15. Debbarma, Mohan (2014). "Environmental Philosophy in the *Borok* Community of Tripura", *Nature, Culture and Philosophy*, Saji Vargese (ed), Delhi: Lakshi Publication, p. 239.

REFERENCES

Bhattacharya, Gayatri (1998). *Refugee Rehabilitation and Its Impact on Tripura's Economy*, New Delhi: Western Book Depot.

Bhattacharjee, Suchintya (1992). *From Jhuming to Tapping*, Agartala: Director of Welfare for Scheduled Tribes and Scheduled Castes.

Debbarma, Debapriya (1983). *Treatise on the Traditional Social Institutions of the Tripura Community*, Director of Welfare for Scheduled Tribes and Scheduled Castes, Agartala.

Debbarma, Mohan (2014), "Environmental Philosophy in the *Borok* Community of Tripura", *Nature, Culture and Philosophy*, Saji Vargese (ed), Delhi: Lakshi Publication.

Ganguly, J.B. (1968). *Economic Problems of Jhumias of Tripura: A Socio-economic Study of the System of Shifting Cultivation of Transition*, Kolkata: Bookland.

Jena, K.N. and Tripura, Babu Dhan (2009). *Changing Face of Tripura Tribe*, New Delhi: Abhijeet Publications.

Jena, K.N. and Tripura, Babu Dhan (2009), *Life and Customary Laws of Tripura Tribe*, New Delhi: Abhijeet Publications.

Sen, Sipra (1993). *Tribe of Tripura: Description, Ethnology and Bibliography*, New Delhi: Gyan Publishing House.

16

Folklore and Nature in the Context of the Khasis: A Look at Some of the Cave Lores

Finley Syngai

Introduction

There has always been a close relationship between humans and nature. Studies about the evolution of culture, social thought and religion in different parts of the world reflect on this close-knit relationship between the two. 'Folklore' literally understood as "the lore of the people" particularly explains the relationship between humans and nature. The Oxford Dictionary defines it as "The body of traditions and knowledge on a subject or held by a particular group". Further lore is a set of doctrines or precepts. It is the accumulated knowledge that has formed a set of communications and dialogue "within a group reflecting the environmental and socio-cultural life and ethno-history" (Sen, 2005). Folklore consist of legends, music, oral history, jokes, proverbs, popular beliefs, fairy tales, stories or customs included in the tradition of culture, sub-cultures or group. It also includes the set of practices through which those expressive genres are shared. The word was coined by William Thoms, in *The Athenaeum*, in 1846 to refer to the shared understanding of the world, the physical setting and the social context of a people. As Botkin remarks in the introduction to his work *A Treasury of American Folklore* (1944), "What makes a thing a folklore is not only that you have heard it before yet want to hear it again, because it is different but also because you want to tell it again because it is anybody's property". It is the lore of the people. Folklore studies enables, as Dunde puts it, the "folklorist to elicit the meaning" out of the folk, about the 'how', 'what' and 'why' behind things people do.

In North-East India folklore continues to hold a special place in the social life of the people. North-East India is extremely rich in the matter of its folklore heritage. For this fact, it has often been referred to as the anthropologist's paradise. With equal justification, if not greater, the region could be called the folklorist's paradise also. The stock of folklore material, both verbal and non-verbal, here, is virtually inexhaustible. Here is where tradition had held its sway till very recent times. Due to heavy influx, spread of education, development and modernization, today, there is a new concern with folklore in the entire region.

Folklore in the Life of the Khasis

Carved out of Assam in 1973, Meghalaya's population belong to three major tribal groups – the Garos, the Khasis and the Pnars. Other tribal groups and non-tribal groups are also found in the state. Although the Khasis, the Jaintias and Garos have come under the influence of Christianity, these groups have not dispensed with their traditional narratives, though connected to traditional beliefs and rituals. In very recent times, there has developed a conscious tendency not only to preserve this folklore heritage, but also to recover what is apparently in the process of being lost.

The Khasis have many attractive legends, myths or folklore. Some legends connected with hills, streams, falls and caves are most touching. The folklore are not only packed with information about the past, but reflect the great connection or knowledge about nature. The unity of nature and culture, as expressed in the form of myths or folklore, is the cardinal point of Khasi thought. The Khasi learn from nature as nature is his natural text (*Ka Sla Khyndew*). The close connection of man to nature is seen through the uncontested belief in the use of a common language by man, animals, birds, trees, rocks, rivers etc. as reflected in their folklore. The world of man cannot be separated from the world of plants or animals. Thus, the Khasi folklore and myths must not be interpreted in the light of scientific thought, but in the light of the Khasi worldview. The name used to refer to a bird or a beast, a plant or a flower is particularly interesting as the term would also depict the true nature of the bird or the beast, plant or flower. In fact, the flora and fauna are named according to their character. As pointed by Ramsiej (2006), *Kynbat Sam Thiah* or sleeping herb gets its name from its movement of which is similar to that of the sun. If the leaves are closed, the farmer understand it is time to go home. Similarly, the leaves of the rubber tree known as *Sla Jri* serve as indicators of the movement of fishes in different streams in the Khasi hills.

Folklore has always been a great part of the Khasi way of life. Stories were told 'round the hearth' (*Sawdong ka Lyngwiar Dpei*) about life, work, migration, family history, morals etc. on a daily basis. Tales and narratives ranging from flowers and trees to rivers and hills, to food and fruits, to animals and birds, explaining the social mores and norms were told to the younger generation with warning to not deviate from the standard pattern of behaviour and life. For example, the tale of '*Ka Tiew Lalyngi Pep Shad*' talks of a maiden preoccupied with herself image, that she missed the great community dance. This warns people not to dwindle with time and lose out on the essential things in life. There are tales of '*Tiew Mationg*', '*Tiew Dohmaw*', '*Ka Iew Luri Lura*', '*Ka Lyn Byrku Sang Khyndew*' etc., which are all told in a similar manner. There exist interesting stories about mountain peaks, caves and rivers. The story or narrative of *U Manik Raitong* linked with the hill *Lum Raitong* is a fine specimen. Khasi folklore speaks volume of the intimate man-nature relationship. The folklore had enabled the Khasis to respect nature as much of their existence and well-being depends on it. Nature is not just the surrounding; it is so much more. This brings to our attention the Khasis' great knowledge about the underworld as well.

A Look at Some of the Cave Lore of the Khasis

Cave lore or simply stories or narratives related to caves, are not only interesting, but informative as these explain the origins of clans, rites, the end of demons and so on. As Brian Kharpran (2013), writes, "the mysteries of caves have for so long been beyond the imagination of man. They were believed to be the places where the spirits of the dead resided, either as evil or harmful ghosts or as friendly spooks . . . for most, caves were a dread, as they were the gateway to hell. Early man used caves as refuge from the ravages of nature and wild animals, and later, in their desire to protect their culture and religion, they used the deeper part of the cave to preserve their beliefs in drawings, which have survived to this day. But all in all, caves have always been an enigma". For the Khasis, right from the days of old, there were stories or narratives that were told about the caves found in the different parts of the Khasi hills.

Caves are the greatest mystery of the planet earth. The cave is the only place on earth with secret chambers and passages not yet discovered and it is the only place that encompasses the mysteries of the unknown. Studies of caves should interest 'us' (the Khasis) more as much about our origin and past can be known from the things found inside a cave. Stories about our ancestors loom round caves. Let us take a quick look at some of the cave lore of the Khasis.

The Cave Lore on Krem Marai — The famous story or legend about the damsel who lived in a cave called Marai, which is situated near Pomlakrai village, "at the source of the river Umiam Mawphlang or Umiew" (Ramsiej, 2006: 28 & 29) is significant not just because it explains the origin of two very important clans – Syiem Mylliem and Syiem Khyrim, but also because the legend talks of the sacredness of the cave wherein dwells the daughter of *U Lei Shyllong* or Shillong God. The damsel was called *Ka Pahsyntiew* or the 'the flower lured one' obviously because the man named *U Sati Mylliemngap* from Nongjri Kongor village who had settled in Nongkseh village lured her by showing her a flower called *U Tiew Jalyngkteng*. After giving birth to sons and daughters, she returned to the cave never to come out again. The children thus became the Syiems and they were called *Ki Syiem Blei* or the god kings of Shillong (*ibid.*). This cave lore provides an explanation as to why the greater part of Shillong is under the rule of the Syiem Mylliem and Syiem Khyrim respectively.

The Cave Lore on Krem Lamet Latang — Krem Lamet Latang, aka Sun's Cave, Krem Sngi (Cave of the Dead), has a story behind it that explains why the sun (and the moon in some versions) hid in a cave when sin (ka pap ka sang) crept and covered the earth, after the golden ladder was broken. Consequently, darkness fell over the earth. All missions to bring the sun back failed but the modest cock succeeded.

The core of the Khasi faith is said to dwell on this myth and Dr. Helen Giri suggest in her book, "Khasi under British Rule", that Khasi religion was born with the cock as the mediator, out of the story about a cave. It becomes a part of the important ritual when a cock is sacrificed during the funeral ceremony. Behind this myth lies the restoration of the ideal relationship between humans and nature. P.R. Kyndiah in "Meghalaya – Yesterday and Today", narrates another version with a lot of action. "The sun fled to her house in the cave covered by big green leaves, leaving the world in total darkness. Confusion reigned and . . . man convened a mass gathering of all living creatures". The assembly selected the elephant but because the way to the cave was narrow but the elephant was huge, the hornbill was selected instead. The hornbill being captivated by the beauty of the sun approached her romantically instead. This enraged the sun to put a curse on the hornbill to henceforth carry asthmatic coughing. However, the cock was selected and it succeeded to bring the sun out of the cave, because it approached the sun with humility.

The Cave Lore on Krem Thlen — In the olden days it was believed that there was a market in the village of *Langhiang Kongkhen* and people going to the market were to pass by *Rangjyrteh* where there was a cave tenanted by a gigantic serpent locally known as *U Thlen*. It was believed that when people passed by the area they would be swallowed by the *Thlen*. The *Thlen* did this according to an order he had received to swallow half of the number of the people passing by the cave. This continued till it was realized by the people that the number of people that had gone missing was increasing. The people then thought of a way to slain the *Thlen*, and counselled a man called *U Suidnoh* to befriend the *Thlen*. However, one day *Suidnoh*, tricked the *Thlen* by asking it to open its mouth wide so that it can enjoy a taste of the famous Sohra pork, as the *Thlen* trustingly opened his mouth *Suidnoh* dropped red hot iron into the mouth of *Thlen*, who died shortly after a struggle. Thus, the cave is known as *Krem Thlen* and the waterfall closeby is known as *Ka Kshiad Dain Thlen*.

The Cave Lore Arlad Umri Jynsuh — This lore is about a contest between two young men from two villages, trying to settle once and for all to which village, Wahlong or Mawlong, the river would belong to. On deliberation it was decided that the two villages would be represented by their best youth. The competition was for the young men to weave a basket under water. The man from Mawlong explored the river before the competition and discovered a cave in the depths of the river. On the day of the contest he was asked to start first, he dived into the water with the necessary materials, made his way into the cave and surfaced and sat on a huge rock to weave the basket in just two hours. The young man from Wahlong could not complete the task and henceforth his village lost the right to the river.

Similarly, there are other caves associated with beautiful cave lore. There are other caves with their own lore and significance. *Krem Kynthon*, for example derived its name from a farmer named, Rai Kynton, who grew paddy and potato and owns a good number of orange trees. It is said that, every year when the oranges are ripened he would pluck the oranges and drop a basket full of oranges down a hole near his house.

He would then sell the rest of the oranges in the market. When the locals asked him about he would say that he was giving the oranges to his children. It was also told that the orange peels were found in the valley.

There is another lore connected to another cave in the Khasi hills, *Krem U Rang Ka Thei*. This cave is situated in *Khadar Shnong*, a village near Cherrapunjee. The cave derived its name from an archer who managed to kill the kite, believed to be the evil spirit that caused plague, death and mysery. The archer killed the kite in the cave with a flaming arrow, as directed by god himself.

The cave lore about *Krem Dam* in Mawsynram village is particularly interesting. Many generations ago, a motley group from the village decided to brave the unknown darkness of the cave, which like a monster swallows up all the water from a river running right into its mouth. Men, women and babies entered the cave with tourch in hand while singing and chanting to drive out evil spirits. As they moved deep into the dark bowels of the cave, "they marvelled at the beautiful formations they came across, shelves and beds, pots and pans, tables and chairs, curtains adoring the ceiling, and breasts swollen and dripping, all patiently formed in rocks. They gaped in awe at the splendour unfolding them. In excitement, some of them out of sheer habit spat red spittle of *kwai* they were chewing constantly. Late that night . . . they dreamt of evil fairies who told them that the entrance of the cave would have been closed and that they were doomed to die inside, for the desecration they had committed inside the cave. What saved them was the presence of some good people in the group whose hearts, filled with reverence and humility, and respected the sanctity of the subterranean world (Kharpran, 2013). It is believed to this day that dark spirits dwell in the *Krem Dam* while good spirits in the famous *Krem Mawjymbuin*.

Thus, legends are born in a cave and over a period of time the legends and stories would have grown and refined themselves. Stories are interesting as they add colour and an aura of mystery to the cave. Perhaps this is the wise move made by our ancestors to preserve the sanctity of the cave. Caves are treated as forbidden places or evil holes thick with the devil's breath. The book "The Long Drop" by Brian Kharpran, expresses the author's longing to explore *Krem Khung* in the *Shnongrim* ridge of Meghalaya. The Shnongrim Ridge proved to be one of the richest areas in terms of cave density in the whole country and was the location of the longest cave system in the Indian sub-continent. The book reflects a lot on the daily activity and attitude of the people. The *Khung*, the animal, the cave is named after had become a myth, a legend – an animal of the past, an animal hunted to extinction. The locals of *Shnongrim*, dare not enter the *Krem Khung* for fear of a demon who resides inside the cave. Kharpran said that this is quiet logical that if there are frequent visitors into the cave it would suffer some damage. However, on expedition, it was realized that the cave drops several metres down; thus if unequipped and untrained, the caver may not be able to come out alive. Thus, the people in the past created a story about a demon residing inside the cave to prevent people from entering the cave and to avoid death.

Caves and Environment Conservation

In today's world environment has taken the central place in conversations, newspapers, magazines, television and so on. It is now understood that our very existence depends on how we treat the environment. "The effects of global warming are telling on us today. If we do not check it, the Arctic ice will melt and the ocean level will rise by many meters to submerge large areas of the earth. Already, the Arctic ice has started to melt. The earth's temperature will soar and life on earth will disappear" (Kharpran, 2013). Thus, nature must be respected as nature provides for man almost all the things he needs. People must be sensitized to their surrounding and the eco-system and strive towards their protection and conservation.

In the area of cave conservation and protection, the works of a local hero named Brian D. Kharpran Daly deserves special mention. Brian Kharpran came forward to create awareness about the importance of caves. He not only wrote books that contain valuable information about cave-life and cave formation, but also started the Meghalaya Adventurers' Association in 1992, and has been exploring caves ever since and knows this underground world intimately. In 2002, he was awarded the prestigious Tenzing Norgay National Award by the Government of India for his work in discovering and documenting caves. In an interview with Brian D. Kharpran, on the evening of 25th May, 2015, about the source of his inspiration or attraction for caves, he said that the inspiration came to him when he was just a child. He further informed that the vast limestone deposits in the state meant existence of many caves. Therefore, the urge to discover, to explore, to document and to map them dawned. With a sense of satisfaction announced that the whereabouts of 300 caves are known in Meghalaya, with about 825 of them explored to yield a total length of mapped cave passage to 376 kilometres. Brian

Kharpran, aims at creating an awareness of what a cave is, its formation, its intrinsic scientific value, the dangers it poses to the novice and the importance of its conservation. He expressed deep concern about the growing pollution and rampant mining that goes on in the state. "Do you see those pieces of wood and plastic bags hanging up there in the ceiling?" (Kharpran, 2013: 12).

Caves are places of unimaginable beauty. Each cave is unique in terms of beauty, fauna and length. Caves are museums of knowledge about the history of the earth – its rocks, climates and years of existence. Caves constitute a valuable scientific resource and are regarded as natural museums in which evidence of past climate, past geomorphic processes, past vegetation, past animals, and past people will be found by those who are persistent and know how to read the pages of earth history displayed for them (Gillieson). For example, an active stalactite when halved along its length will reveal rings, like a tree trunk, made of different width, colour and density. These are the records of the monsoon in an area for thousands of years. Similarly, knowledge is gathered about the temperature of the formation of the cave. Caves are of great interest and importance to a geologist, a hydrologist, an anthropologist, an archaeologist, a botanist, a zoologist, a folklorist and so on. Caves are very important and at the same time caves are very fragile and vulnerable. Anything that affect the surface of the earth will directly affect caves. Quarrying for example is a great enemy of a cave as this activity can simply remove it. Another danger comes from the mining activities near the vicinity of a cave, particularly in an area that receives heavy rainfall, will lead to weathering of rocks, thereby leading to leaching metals that cause acidic drainage, which is poisonous to aquatic life inside and outside of a cave. Deforestation will also lead to the sedimentation inside a cave that can block the free passage of air inside a cave. Caves must be conserved as they are of great value.

Caves are the greatest mystery of the planet earth. The cave is the only place on earth with secret chambers and passages not yet discovered. It is the only place that encompasses the mysteries of the unknown. Studies of caves should interest 'us' more he said, as much about our origin and past can be known from the things found inside a cave. Stories about our ancestors loom round caves.

Talking about pollution and its impact on the caves, he said, the need of the hour, is to sensitize the people about the importance of caves not only in terms of the associated myths and legends but most importantly in terms of their significance in the environment. Talking about his opinion on the N.G.T., he said his concern is more on how mining is done. Coal mining goes deep and the extracted coal rich in sulphur cannot be left on the land surface without care. When rain sets in sulphur is leached from it and it becomes carbonic. When this water enters or seeps into a cave it will cause destruction and affect the evolution of flora and fauna that are very specific in a cave. It must be brought to our attention here that each cave is endowed with a special array of life formation in terms of its flora and fauna. The fishes inside a cave found in minute numbers would be most affected. Again reclamation of land or the diversion of rivers that goes inside a cave will affect the flow of air that will adversely affect cave life. Thus, mining should be properly planned to not affect the delicate nature of a cave. The master cave system as found in Siju cave must be protected. The beauty of a cave must be preserved for the next generation to see. Krem Labit in Kseh village is the most appealing as there is a big entrance to it; swimming into the dark is a ceiling lined up with bats and then, there are these dry passages and beautiful formation of gypsum flowers protruding out of the walls. There are black crystal flowers over a stretch of eight kilometres. Krem Synrang, in his description is the most beautiful cave in the region. The chamber is clean and pure white, with crystal pools inside and adorned with cave pearls. He said he is glad the chamber is deep inside the cave so that people will not be able to disturb it or more so, extract the pearls and crystals; and hence destroy the cave forever.

Indeed as in other parts of the world, the pressure of development, increase in population, increasing migration, increasing pollution to name a few are all adding to the increasing need to protect and conserve the environment, and particularly the caves. It is in this context that one learns to acknowledge of the people in the past who wisely created an interesting way, that has lead to cave preservation in the form of cave stories or cave lore. As mentioned earlier the lore on caves truly reflect the knowledge the people have about the environment in its totality. Thus, many of the caves in the Khasi hills have been refrained from damage because of the existing lore that the people know and share in common.

Eco-criticism

Clearly, the advent of the Industrial Revolution ushered changes in various aspects of life. Society 'evolved' from work by hand to work by machines, from agricultural economy to factory based economy, from leisure and work to pressure and work etc. Thus, society changed and the total 'environment' also changed. Concern about the environment (physical and social) trickled in with a built-in paradox in itself. The paradox lies in a powerful and honest effort to resist and curb such changes (as found in the works of poets, novelists, historians, psychologists, sociologists); and in the realization that the natural resources should be tapped and used, to ease human problems and hence improve human welfare. This paradox goes to the very heart of issues in environmental and ecological studies. It resulted in the emergence of various disciplines and schools of thought or theoretical perspective.

Eugene Odum, who coined the word 'Ecosystem', offers cogent and logical arguments for the emergence of 'Ecology' as an integrated discipline (Odum, 1977). He emphasized that the concept of ecology is frequently misused as synonymous with environment. As he uses ecology, it includes man as a part of rather apart from his natural surroundings (Krasner, 1980: 20). Odum emphasized on the interaction between man and his environment in holistic systems. The development of ecology, the science of relationship between organisms and their environment, is also called 'Bionomics' (*American Heritage Dictionary*). The growth of Human Ecology analyse community structures and social areas, and Cultural Ecology emphasize on the impact of culture as an entity on the usage and interpretation of physical environment. The recent interest in conservation and the 'quality of life' has become the basis for the ecology movement. Resource depletion, pollution and population growth are examples of the kinds of environmental concerns which increasingly come to the fore in research. Moos credits Rachel Carson's "Silent Spring" (1962), as being of seminal importance in arousing public awareness of the consequence of environmental pollution.

Eco-criticism explores the ways in which the relationship between humans and the environment is imagined and portrayed. It is inspired by, but also critical of, modern environmental movements. It is also the most exciting recent developments in literary and cultural studies.

Eco-criticism is the study of the relationship between literature and the physical environment. It is generally agreed that eco-criticism begins with 'A Fable for Tomorrow', in Rachel Carson's 'Silent Spring' (1962). The striking quality of the book is that it is conceived and presented as a fairy tale. It opens with the words, "There was once a town in the heart of America where all life seemed to live in harmony with its surroundings". Carson presented a perfect picture of 'harmony' of humanity and nature, of essential changelessness. Then, every element of the rural idyll is torn apart by some agent of change, the mystery of which is emphasized by the use of both natural and supernatural terminology of 'malady' and 'spell' (Garrard, 2004: 1-2). The 'Silent Spring' alludes the loss of birdsong, which in a way sounds the alarm of an unnatural nature; an environmental apocalypse.

The real culprits, according to Carson, were the new organic pesticides such as DDT, aldrin and dialdrin that had been introduced after Second World War (Green Revolution), and had resulted in successful pest and insect control. Silent Spring marshalled an impressive array of scientific evidence to show that this very success constituted a serious threat both to wildlife and to human health, confronting the utopian claims of agricultural scientists on their own ground. This increased public awareness of pesticide pollution, firmer state regulation and development of less harmful agricultural chemicals. Such analysis is referred to as 'eco-criticism'.

Eco-criticism is unique amongst contemporary literary and cultural theories because of its close link with ecology. Eco-critics generally tie their cultural analysis explicitly to a 'green' moral and political agenda. Eco-criticism seeks to evaluate text and ideas in terms of their coherence and usefulness as responses to environmental crisis (Kerridge, 1998: 5). Eco-critics may not be able to contribute to debates about problems in ecology, but can certainly create understanding on such problems. One eco-critical way of reading is to see contributions to environmental debate as rhetoric (*ibid.*). Thus, eco-critics transgress disciplinary boundaries and develop their own 'Ecological Literacy' as far as possible. This will enable the treatment of the changes around as the production, reproduction and transformation of large scale metaphors.

Indeed, eco-criticism is the study of the relationship of the human and the non-human throughout human cultural history and entailing critical analysis of the term 'human' itself. Carson had to investigate a problem in ecology with the help of wildlife biologists and environmental toxicologists, in order to show that DDT was present in the environment in amounts toxic to wildlife, but Silent Spring took a cultural not a scientific stand to present the problem at hand. The great achievement of the book was to turn an ecological problem into a widely perceived problem that was then contested politically and legally. Thus, eco-criticism can help to define, explore and even resolve ecological problems in the wider sense.

The 'Silent Spring' produced four distinct eco-philosophical movements or approaches. Each approach understands environmental crisis in its own way, with specific literary or cultural affinity and aversions. These are – Cornucopia, Environmentalism, Deep Ecology and Eco-feminism.

Cornucopia literally means "an abundant supply of good things". It is a symbol of plenty consisting of a goat's horn overflowing with flowers, fruit and corn. It is derived from the Latin word 'Cornucopia' meaning 'horn of plenty' (*Compact Oxford English Dictionary*, 2008). This school of thought puts a lot of faith in technology to improve human living conditions and to solve environmental problems.

Environmentalists are conscious about natural resource scarcity, pollution, global warming, but would like to maintain or improve their standard of living. Activism range from recycling bottles and buying organic food to major commitment to conservation activity. Environmentalism is widespread and very powerful.

Deep ecology encourages a more extreme radical approach to the environment. Its influence spread beyond the academic circles, inspiring many activists in organizations such as Friends of the Earth, Earth First and Sea Shepherd. The 'poet laureate' of deep ecology is Gary Snyder, a folklorist and a poet. Its philosophical guru is Arne Naess. Deep Ecology criticize the 'shallow' approach of the environmentalists, and at once demand recognition of intrinsic value in nature. Therefore, it tries to revive the notion that the world is sacred. Gary Snyder affirms this in his book "*The Wild and the Sacred*". The treatment of the world as a sacred place is at the core of majority religions. Therefore, the duty of all people is to protect it, though earth worship was found in oral religions. Responsibility of nurturing nature rest upon all of humankind. Thus, deep ecology has proceeded from, and fedback to Eastern religions and mysticism, ancient Christianity, shamanistic and other primal religions.

Eco-feminism associate women with nature. Radical eco-feminists revive goddess worship as they claim only women can feel the importance of the environment more than men. Sharon Dubiago asserts that "ecology consciousness is traditional woman consciousness" (Garrard, 2004: 24). Charlene Spretnak, similarly, highlights a special kind of women's spirituality in female biology similar to nature. These arguments intend to change the hierarchy of men over women, by showing instead how gender is culturally constructed. In fact, the patriarchal construct of femininity limits and hold back women from many activities. Thus, eco-feminists, critique gender role as structured in a patriarchal society. They cherish the 'Mother Earth' concept, and criticize discrimination and oppression on grounds of race, sexual orientation, class, species and gender.

Khasi Understanding of Nature

Tribes live in harmony with nature and maintain symbiotic relationship with it (Reddy, 2002: 34). The close and symbiotic relationship that these primitive people have with the natural world is build up over generations. The tribal's ecological way of knowing nature are necessarily participatory (*ibid.*: 83).The Khasis look at nature as a living and growing organism suffused with life and intelligence. Nature stands for the totality of things including man. The legends of the Khasis relate the continuous dialogue between man and natural forces. The relationship is one of give and take and what holds good for nature holds also for man (Ramsiej, 2006). The Khasis respect their mountain peaks, rivers, sacred forests, monoliths, trees and caves because they are sacred. It is, therefore, impossible to break or separate man from nature. He does not ascribe to himself a unique and a privileged place in the scale of nature. Life possesses the same dignity in its highest and lowest forms. Men and animals, animals and plants are all on the same level. A Khasi lives so close with his Mother Nature and nature lives in him. It is like a scripture or a text and a treasure or library of all knowledge. As observed by Mawrie in *The Khasi Milieu*, nature teaches and guides man in his day to day life. Nature provide for man a sense of time, it helps man to know the changes in weather and seasons. He looks for signs in nature to understand the coming of rain or shower, to understand the time for sowing seeds. He looks for herbs in nature to cure himself from ailments of various types. The unity of nature and culture is strongly expressed in Khasi thought.

It is interesting to note that the Khasis have always understood the value and importance of nature. Man is an agent of nature and must, therefore, live, understand and behave according to the laws of nature and most of all respect mother nature. As discussed above, to accomplish this, the Khasi devised a way to teach the young and old of the importance of nature. The teachings come in the forms of folklore or narratives, or legends that are coated with morals. The Khasi always consider nature as the source of life, fertility and continuity and hence they address nature as *Mei Ramiew* or *Mei Mariang*. Nature contributes to Khasi thought and in a significant way to the Khasi way of life. All activities revolve round nature from farming, to music, dance, festivals, to stories and lore. In fact, Khasi religion though not naturalistic, teaches man to respect man, to make man understand that man is a creature of God as other things around him is also a creation of God.

To conclude, it can be said that the Khasis are very sensitive to nature. It is the source of life and existence and hence must be respected. They created different ways to make people aware of the great value of nature. This concern about nature is as old as Khasi thought itself. In fact, the ideas and concerns reflected in Eco-criticism have always been a part of the Khasi worldview or simply understanding. The narratives or folklore have been instrumental in creating awareness about nature, perhaps the way Silent Spring has managed to usher great concerns on the environment and its preservation. Myths about *U Thlen* for instance warns people of the evils of having a Cornucopean attitude towards excessive wealth accumulation, which proves destructive to man and nature. Similarly, the beliefs cherished by Deep Ecologists has always found expression in the Khasi treatment of nature as something sacred. Lastly, Eco-Feminist's treatment of nature as the source of life and fertility, and the Mother Earth or Mother Nature concept is also strong among the Khasis who consider nature as *Mei Mariang* or *Mei Ramiew*. Therefore, it is clear that nature is very close to a Khasi and it lives in him. The attachment is so strong that a separation of the two becomes an impossibility.

BIBLIOGRAPHY

Barr, John (1969). *Derelict Britain*, Penguin Books, England.

Daly, Kharpran, B.D. (2013). *Caves for the Uninitiated*, Strategic Book Publishing and Rights Company.

Daly, Kharpran, B.D. (2014). *The White Climb*, Partridge India.

Daly. Kharpran, B.D. (2014). *The Long Drop*, Partridge India.

Datta, B., Sarma, N.C., Das, P.C. (Ed), *A Handbook of Folklore Material of North-East India*, ABILAC, Ambikagiri Nagar, Guwahati. Assam.

Garrard, G. (2004). *Eco-criticism*, Routledge, Milton Park, Abingdon, Oxfordshire.

Handoo, Jawaharlal (Ed.) (1998). *Folklore in Modern India*, CIIL, Mysore, India.

Laloo, Minimon (2007). *Ka Kolshor Baiạr u Hynniew Trep* (Ka Iitkhmih u Mynnor), Laloo Brothers, Mission Compound.

Lyngdoh, M.P.R. (1991). *The Festivals in the History and Culture of the Khasis*, New Delhi: Vikas Publishing House Pvt. Ltd.

Lyngdoh, Homiwell (2013). *Ka Niam Khasi*, Shillong: Ri Khasi Book Agency.

Mawrie, B.L. (2014). *Khasi Ethics*, Shillong: Vendrame Institute Publications.

Mawrie, H.O. (1981). *The Khasi Milieu*, New Delhi: Concept Publishing Company Pvt. Ltd.

Mawrie, H.O. (1973). *U Khasi bad la ka Niam*, Shillong: Ri Khasi Press.

Moos, R.H. and Humphrey, B. (1974). *Group Environment Scale*, Palo Alto, Calif Consulting Psychologist Press.

Moos, R.H. and Insel, P. (ed.) (1974). *Issues in Social Ecology: Human Milieus*, Palo Alto, Calif. National Press Books.

Odum, E.P. (1977). *The Emergence of Ecology as a New Integrative Discipline*, New York: Academic Press.

Ramsiej, K.D. (2006). *Khasi Philosophy of Nature: Seven Huts*, Shillong: Socio-Educational Association, Mawlai Umthlong.

Reddy, K.V. (ed.) (2002). *Cultural Ecology of Indian Tribes*, Delhi: Raj Publications.

17

Revitalizing Traditional Knowledge in North-East India by Combining Conservation, Research and Entrepreneurship: Living Root Bridge Case Study

Sanjeev Shankar

Revitalizing Traditional Knowledge in North-East India

North-East India represents an eclectic mix of cultures and beliefs, which show immense wisdom in their understanding of the environment and its contribution to a society's survival and well-being. As a collective society, which is gradually transforming and integrating with the mainstream, it presents a unique context for studying ancient practices and developing ingenious hybrid solutions, which are inclusive, authentic and sustainable. By integrating different viewpoints and balancing individual needs with the needs of a community, the author proposes a mergence of ecological and economic concerns, and developing solutions together with local communities in a non-hierarchical, symbiotic, participatory and immersive manner. Here traditional knowledge and contemporary thought will be given equal value, and solutions will develop through rigorous field tests and objective analysis. The author proposes a three-pronged approach to work with traditional knowledge, which focuses on (1) conservation, (2) scientific research and development, and (3) grassroot entrepreneurship. Connecting these domains (Figure 17.1) will create a fertile ground for traditional knowledge to grow and adapt in response to changing realities at a local, regional and global level. Adopting these fields as essential pillars for working with cultural heritage and ancient wisdom will lead to novel solutions, which are resilient and representative of different sections of society, and allow transmission of this knowledge to younger practitioners.

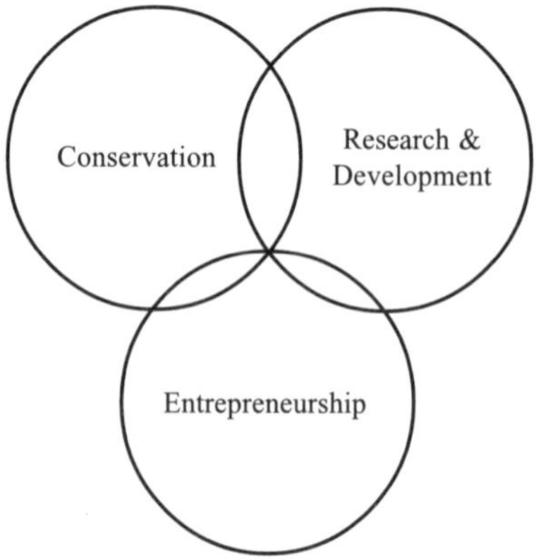

Figure 17.1: Pillars for Working with Traditional Knowledge.

Source: Author's study.

Living Root Bridges

Indigenous Khasi tribes (Fig. 17.2) of Meghalaya are viewed to be an offshoot of the Mon-khmer branch of the Austro-Asiatic stock and are believed to be due remnants of the first Mongolian overflow into India. Demonstrating a high degree of self-sufficiency, which in part is owing to their remote location and distinctive environment, Khasis have developed numerous sustainable practices based on collective and planned cooperation. Through an elaborate attempt to balance individual needs with the needs of the larger community, they have nurtured an

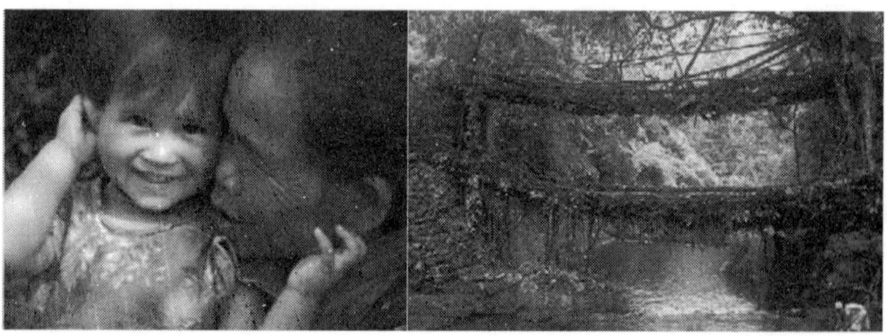

Figure 17.2: Khasi Tribes and Living Root Bridges of Meghalaya.

Source: Author's study.

ecosystem, which acknowledges the interdependent and interconnected nature of all life. Relevant systems worth highlighting include the practice of sacred groves, coherent classification of land and forests, an independent village-durbar based system of governance, an eco-theandric vision of reality, laws of inheritance and succession, and laws of consanguinity and kinship. These ecology related management practices are an indispensable part of Khasi life and critically contribute to resiliency within the community.

Khasi tribes' intimate and symbiotic relationship with their natural environment is epitomized in the living root bridges of Meghalaya (Fig. 17.2). Locally known as *jing kieng jri*, this indigenous technology uses the aerial roots of *Ficus elastica* to grow bridges ranging in spans from 15 feet to 250 feet over a time period of 15 to 30 years. Situated in heavily forested and wet places, which are prone to flash floods and storm surges, the underlying growth process of these plant-based systems involves nurturing the aerial roots of *Ficus elastica*, and guiding them across rivers and streams. The supporting horticultural technique uses locally available plant based material, local tools, and community based expertise. Key advantages of the living root bridges include low cost, high robustness, remedial impact on surrounding environment, carbon sequestration, community involvement in the growth process, progressive increase in load bearing capacity, and support for other plant and animal systems. Despite these attributes living root bridges are being replaced by inappropriate solutions owing to its long growth phase, low safety during initial growth stage, need for appropriate growth conditions (soil, water, sunlight and nutrients), and a sensitized community support for nurturing its growth and maintenance. In addition, the nexus of poverty, population explosion and environmental degradation in North-East India, poses complex challenges from a development viewpoint and highlights the need for adapting this traditional knowledge. The author proposes a transdisciplinary socio-scientific-entrepreneurial research framework to understand, improve and adapt these bridges for local needs. By aligning local economy and rural development requirements with *Ficus* ecology, the author proposes to use living root bridges as a source of food, medicine, nutrition, and other plant and animal based allied products for all-round development and well-being.

Conservation

Preservation and protection is the first step in a sequence of successive measures required for working with traditional knowledge. This creates a reference for all subsequent action. In the context of North-East India, this stage assumes critical importance owing to rapid changes occurring across all domains of the society. With specific reference to ecology related practices, conservation assumes high priority owing to the inherent nature of a living ecosystem. Once lost or damaged, remediating and restoring such ecosystems to their original performance is a challenge. Old growth forests and sacred groves are a key reference in this context. In the case of the living root bridges, the author proposes a state-wide mapping, ethno-botanical study to reveal the current state of these bridges, their overall performance, and the local communities relation with these structures. This will lay a foundation for evaluating the scale of the conservation challenge and facilitate an appropriate local strategy. It will also reveal critical insights for scientific research and development.

Research and Development

Working with traditional knowledge requires a precise understanding and assessment of the process and product. Well-informed immersive research has the potential to alter the way we 'see' and 'value' a process or a technology, and create superior solutions, which lead to all-round development. The team has to find a critical balance between the 'old' and the 'new', and celebrate aspects, which demonstrate potential, and in turn, therefore, sacrifice other aspects, which do not contribute to the overall performance. It is essential to work with a sense of innovation and experimentation to develop solutions, which respond to current needs and realities of the society. Another aspect of working in this context is accepting that 'identity' is a rational and dynamic phenomenon. Identities, people, products, perceptions and beliefs are always in a state of flux and this constant change is essential for growth of a system. In order to respond to contemporary concerns and opportunities, and to apply traditional knowledge and skills, the communities must develop a sense of openness and balance between past and present. The author would like to highlight two research and development case studies from Nagaland, which explored a new product range and application for the traditional craft of weaving (Figs. 17.3 and 17.4). Developed with the indigenous Angami tribes of Khonoma and Dimapur, the development process was inspired by local skills, natural materials and indigenous sense of beauty. In the first case (Fig. 17.3), a new range of lamps were developed through experimentation, which focused on developing natural colour based process, and alternate material options (*kona* grass). In the

Basket Weaving in Bamboo & Ratan

Local Grass Developed as Alternative Material Option

Locally available, natural herb based colouring technique developed and applied for new designs to create new community based grassroot entrepreneurial opportunities

Figure 17.3: Natural Fiber Weaving Research and Development, Nagaland.

Source: Author's study.

This project focused on three interconnected research domains: materials, morphology and structure. The material research developed alternative material options with colours using local herbs. The morphology developed new forms using moulds and alternative sources of energy. The structural research leveraged weaving as a performative feature, which provides strength to the overall form and plays a critical role in furniture making. Please see link for further details: http://www.sanjeevshankar.com/cane2.html

second case (Fig. 17.4), traditional timber louver was used as a reference to develop a low cost tropical building technology. A four stage iterative process informed by periodic feedback was followed in both cases as listed below. This process can be used for revitalizing traditional knowledge across all domains, industries and scales.

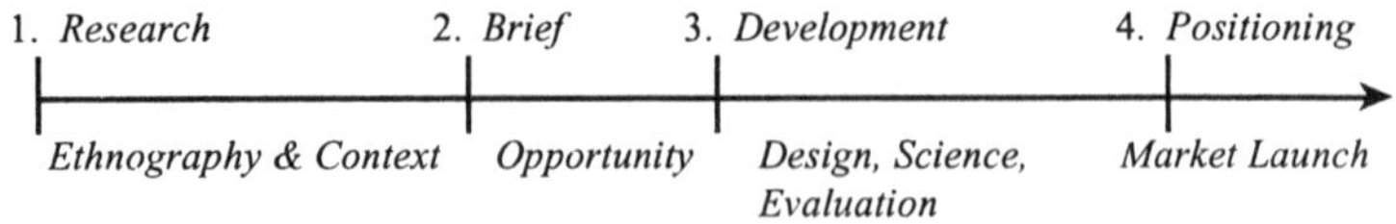

It is important to highlight that revitalizing traditional knowledge requires extraordinary sensitivity and commitment from a social and scientific viewpoint. The underlying processes have evolved over an extended time period and reached a state of equilibrium, and changing this requires absolute dedication. Inspiring local communities to experiment and improvise is faced with stiff resistance in the early stages, where the results are not immediate. This however gradually changes with hands on experimentation, a deepening relationship and blurring of boundaries between the innovator and the maker. With time, the new identity echoes a transformational journey from past to present.

In the case of living root bridges, the author proposes a precise scientific investigation focussed on understanding, improving and propagating the technology. This investigation will require transdisciplinary expertise from the fields of botany, ecology, bridge engineering, material science, environmental science and social science. Key areas for improving the living root bridge technology include: (1) long gestation period of 15 to 30 years; (2) low-safety during early growth stage; (3) need for ideal growth conditions; and (4) need for community support. Testing the horticultural technique for other native plant species, e.g. *Ficus benghalensis*, would offer greater flexibility and choice, and this constitutes a key research area. Findings from this research can lead to an improved living root bridge growth process, which can subsequently inform a state-wide rural connectivity, conservation, education and livelihood project.

Entrepreneurship

Creating ecology based grassroots entrepreneurial opportunities is critical for a self-sustaining development model in North-East India. In the case of the living root bridges, the author proposes aligning local development

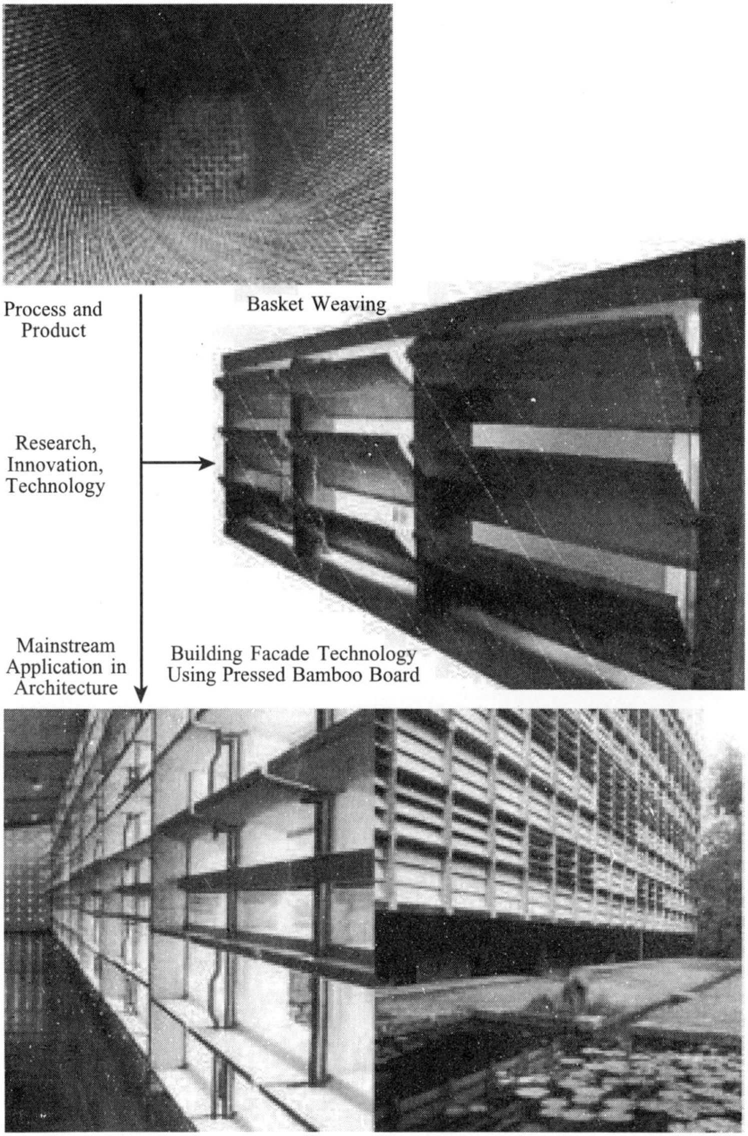

Figure 17.4: Pressed Woven Bamboo Research and Development for Building façade Technology. This building technology development project leveraged traditional basket weaving process and timber louver systems to develop a low-cost building façade technology for mainstream use.

Source: Author's study.

needs with *Ficus* ecology to alleviate the nexus of poverty, population explosion and environmental degradation. Potential adaptation of the bridge technology as a host biome for orchids, mushrooms, other epiphytic plants and food for humans and other biota is proposed as a critical value-addition and livelihood promotion strategy. This will transform the bridge into a source of food, medicine and nutrition for local communities. Other key value-addition interventions include redesign of the bridge to support vehicular movement, potential adaptation of the *Ficus* aerial root inoculation method for developing furniture, lamps and fences, and creating a grassroots 'design and grow' cooperative for providing infrastructure development solutions to other tropical regions.

Figure 17.5: Living Root Bridge as a Source of Food and Nutrition for Humans and Other Biota.

Source: Author's study.

Concluding Remarks

North-East India offers a compelling and unique context for ecology based novel development solutions, which combine traditional wisdom with contemporary thought. Living root bridges demonstrate the intimate relationship between Khasi tribes and the forest, and offer solutions for the future. These structures highlight the essential human value of living with sensitivity and respect for all life, and can inspire a new balance.

Acknowledgements

The author is indebted to the Khasis for hosting him, sharing their knowledge and for guiding him through Meghalaya. He would also like to thank Mr. Aiban Swer, Dr. Rekha M. Shangpliang and Prof. A.G. Rao. Gratitude also goes to other scientists, engineers, architects, entrepreneurs, government representatives and residents who have inspired the author with their support.

REFERENCES

A keystone species is a specie that plays a unique and crucial role in the way an ecosystem functions. *http://education.nationalgeographic.com/encyclopedia/keystone-species/*

An eco-theandric vision of reality refers to a firm belief in God, Nature and Man as one single indivisible entity. Khasis idolize the earth as "mother earth", which is a seamless and indivisible combination of land, forest, rivers and streams. Vidyarthi, L.P. and Rai, B.K. (1976). *Tribal Culture of India*, New Delhi: Concept Publishing Company (P) Ltd., pp. 111-140.

Bareh, H. (1985). *The History and Culture of the Khasi People*, Guwahati: Spectrum Publications, p. 15.

For a detailed elaboration about the living root bridge growth process, research possibilities and future application see Shankar S. (2015). *Living Root Bridges: State of Knowledge, Fundamental Research and Future Application*, International Association for Bridge and Structural Engineering Conference: Providing Solutions to Global Challenges, Geneva, Switzerland, September 23-25. *http://www.sanjeevshankar.com/pdf/Living-Root-Bridges-by-Sanjeev-Shankar-IABSE-Conference-Geneva.pdf*

Khasi Land Tenure system deals with three types of land in Khasi hills, *viz.*, community land (or *Ri Raid*), privately owned land (or *Ri Kynti*) and Government land. Community land is managed and owned by the community and performs the role of an emergency reservoir for the benefit of all people. Shangpliang, R. (2010). *Forest in the Life of Khasis*, New Delhi: Concept Publishing Company (P) Ltd., p. 17.

Native from the Himalayas to Malaysia, Sumatra and Java, *Ficus elastica* (or India rubber tree or India rubber fig) is a broadleaf evergreen shrub or tree that may grow to 50-100' tall in its native habitat. *http://www.missouribotanicalgarden.org/gardens-gardening/your-garden/plant-finder/plant-details/kc/b597/ficus-elastica.aspx*

North-East region is a global hot spot for bio-diversity. Characterized by a varied physical geography, it is marked by distinct orography, heavy monsoon rains and a diverse range of flora and fauna. Shangpliang, R. (2010). *Forest in the Life of Khasis*, New Delhi: Concept Publishing Company (P) Ltd., p. 5.

Resiliency is the ability of a system to adapt to external disturbances and yet remain within critical thresholds. *http://www.stockholmresilience.org/21/research/what-is-resilience.html*

Robustness is used to describe a system that can survive extreme external variations. For a detailed elaboration, see Weinstock, M. (2006). *Self-Organization and the Structural Dynamics of Plants*, AD Emergence: Techniques and Technologies in Morphogenetic Design, Vol. 76, No. 2.

'Sacred groves' (or *Law Kyntang*) are fully developed virgin forest patches from pre-agrarian age. Removal of timber or forest produce from these groves, is prohibited. Tiwari, B.K., *et al.* (1999). *Sacred Forests of Meghalaya*, Regional Centre, National Afforestation and Eco Development Board, NEHU, Shillong, p. 14.

The term "Khasi" means "born of the mother". For a detailed elaboration, see Shangpliang R. (2010). *Forest in the Life of Khasis*, New Delhi, Concept Publishing Company (P) Ltd., p. 1.

18

Production of Traditional Ecological Knowledge: Intersecting Culture and Environment

V. Bijukumar

Introduction

The relationship between culture and environment has found animated discussions in sociological and anthropological literature in the recent past. The cultural ecology, which emerged subsequently, focuses on the relationship of specific cultural features with a group's adaptation to its total habitat. It is viewed that the settlement pattern and agricultural system influence cultural aspects and environmental diversity influence community-based natural resources. The everyday interaction between a cultural community and environment produces new forms of knowledge. Such knowledge production is based on the observation of the environment or the surrounding in which a cultural community inhabits. Since knowledge is produced in cultural context, each culture is considered as a knowledge system which produces, reproduces and rechristens knowledge.

In indigenous community, culture is the basic ingredient of ecological conservation. The conservation of ecology emerged out of community's moral and ethical commitment to non-human entities. In other words, such commitment has the larger context of the traditional ethics of living with nature of indigenous communities and their emotional link with non-human entities such as animals and plants, which they feel as essential for their material survival. In other words, environmental ethics in indigenous societies is built on cultural norms based on the realization that non-human entities are critical for the survival of human beings on earth. Thus, culture establishes a web of relationships between human beings and nature. Marx and Engels believed in the relationship between human beings and nature based on interaction and mutual dependence. Marx, in fact, saw nature as the premise for the development of society and history. He conceived that nature is a part of man's essential being, a confirmation of his existence as a living being, and life world as an interaction between human and nature based on mutual respect. The man-nature relationship is conceived by Marx as a relationship between two living entities, mutually transforming each other. Man and nature do not exist in isolation, and on the contrary establish constant interrelationship between these two. This cordial relationship is disturbed by the development of capitalism, which is based on alienation of both human and nature. The environmental ethics and ecological consciousness of human beings crumbled with the development of the capitalist mode of production.

The interaction between a cultural community and environment produces a new form of knowledge called the Traditional Ecological Knowledge (TEK). The production of TEK is based on the reaffirmation of the environmental ethics of a cultural community, which is not dictated by any legal political authority. On the contrary, TEK is the production of cultural impulses of many civilizations over a period of time and not a spontaneous outgrowth. It is synonymously used as local knowledge, folk knowledge, rural people's knowledge, oral knowledge, applied knowledge and empirical knowledge. It is often described as the experimental knowledge based on observation and transmitted from one community to another. In all these forms, nature is the source of knowledge. Social habits and customs related to culture often establish a relationship between communities, especially the indigenous ones. The preservation of TEK is essential for the survival of larger ecosystems and hence, the growing realization that it can preserve bio-diversity in a society.

The term TEK came into widespread use only in the 1980s though it was associated with primitive origin of human beings. Often used as 'ethnoecology', this establishes ecological interconnections of one ethnic community with nature. TEK is acquired through human beings' direct contact with the environment over years. It is an adaptive process handed down through generations by cultural transitions. TEK is the sum of knowledge, spiritual values and practices passed down from generation to generation about environment. According to Berkes, "the growing interest in traditional ecological knowledge since the 1980s is perhaps indicative of two things: the need for ecological insights from indigenous practices of resource use, and the need to develop a new ecological ethic in part by learning from the wisdom of traditional knowledge holders" (Berkes, 1993: 19). TEK is based on the observation of local ecosystems used for resource conservation. It is the locally developed knowledge aimed at sustaining resource use and environmental conservation. As Berkes argues that "traditional ecological knowledge represents experience acquired over thousands of years of direct human contact with the environment" (Berkes, 1993: 1). The conservation of TEK is now seen as an alternative way of conserving ecology.

In the conception of TEK, the word 'traditional' is often problematic as it is often viewed as opposite to progressive, irrational, conservative, static and old. Tradition is often described as antithetical to modernity, wherein the social attitudes and beliefs are culturally specific, and thereby tradition blocks the quest for change. As tradition gets transmitted from one generation to another, cultural transmission leads to knowledge transmission, and knowledge about the physical world transmitted from one generation to another. It is based on local history, oral narratives, myths etc. Eric Hobsbawm argues that traditions are invented, constructed and formally instituted. According to him, "invented tradition is taken to mean a set of practices, normally governed by overtly or tacitly accepted rules and of a ritual or symbolic nature, which seek to inculcate certain values and norms of behaviour by repetition, which automatically implies continuity with the past" (Hobsbawm, 1983: 1). In '*The Savage Mind*', Levi-Strauss argues that the ancient societies could not have acquired such technological skill as those involved in the making of watertight pots without a curiosity-driven scientific attitude and a desire for knowledge for its own sake. According to him, "the universe is an object of thought at least as much as it is a means of satisfying needs" (Levi-Strauss, 1962: 3).

Traditional societies have certain spiritual relationship. The critical attitude towards traditions is explicated as some traditions are based on exclusion and injustice and certain traditions are the realm struggles for fighting for rectifying it. In this respect, traditionality is opposed to emancipation. In a traditional society, customs act as the basis for order and social control and it gets a bad connotation being based on superstition and dogma. According to Oakeshott, tradition not merely reflects our attachment to the familiar, but also ensures that social institutions work better because they operate in a context of established rules and practices

(Oakeshott, 1962). Shils (1981) says tradition encompasses "anything transmitted or handed down from the past to the present". As tradition is constituted by wide-ranging fields such as institutions – social, political, beliefs and values – and it has a continuity with past, it obstructs rational outlook and thinking of the humanity. Humanity thus tends to be critical of tradition as it retards the path towards progress.

Ecology as the Source of Epistemology

Ecology is the source of the production and reproduction of knowledge in the conception discussed above. TEK is often described as the alternative form of knowledge. The expression of ecological epistemology is related to tradition, culture and practices emerged out of the local experiences, living practices and beliefs of particular community. Local communities are the stakeholders of the knowledge production. The TEK is based on integrated system of knowledge, practices, and beliefs including cosmology (Berkes, Colding and Folk, 2000). The TEK, according to Environmental Information Partnership, based in Ontario, Canada, is: "environmental knowledge that has been gathered by aboriginal peoples who have lived in and observed a particular area for generations, and has been handed down by statements, beliefs, legends, customs, from generation to generation by word of mouth or by practice, based on facts, truths, or principles" (EIP, 1996: i). According to Berkes, TEK is "a cumulative body of knowledge and beliefs handed down through generation by cultural transmission, about the relationship of human beings (including human) with one another and with their environment" (Berkes, 1999: 8). TEK is transmitted from generation to generation as stories, songs, folklore, proverbs, cultural values, beliefs, practices, experiences, rituals, community laws and local knowledge. TEK has certain spiritual and moral dimensions. It is produced only in particular cultural and ecological context and is useful for sustainable environmental development. The importance of TEK is growing in spite of the advancement of scientific knowledge.

Culture and Environment

Culture and religion play an important role in preserving and conserving the nature. Since time immemorial indigenous communities have developed an intimate contact with the natural environment. Nature contributes substantially in the development of culture of the indigenous community. Beginning from the subjective condition of one's position in relation to environment, culture has a vital role in planetary environmental consciousness. The relationship between environment and culture is much discussed in the anthropological literature. Each cultural group possesses a distinct form of ecological knowledge and establishes human-nature relationships. The interaction of cultural community with natural environment produces a new knowledge called the TEK. The conservation and management of environment constitute an important factor in the construction of cultural identity. The relationship between culture and environment is the relationship between biophysical world and the human social world. Culture cannot survive without nature or ecology. Ecological management is essential for the survival of humanity and ecosystem and is constructed on the basis of tradition.

Cultural Diversity and TEK

The source of TEK is cultural diversity. It has been proved that cultural diversity strengthens bio-diversity as different traditions, belief systems and social norms conserve bio-diversity. Any attempt to bring cultural homogenization or constructing monolithic culture can harm the conservation of ecology. Imposition of monolithic culture not only discredits the ecological system, but also endangers the relationship between human beings and nature. The relationship between community and nature is basically reciprocal, which is similar to the one between community members and other beings. TEK is often seen as an expression of cultural diversity, intended for the protection and management of natural resources. Celebrating diversity in a society promotes bio-diversity and the protection of nature and ecosystem.

TEK is based on local observational knowledge of species and also people's understanding about their surroundings. Local social norms and cultural internalization by the groups determine the basis of such knowledge. It is transmitted through oral tradition and first-hand observation. The production of TEK is integrally related to cultural identity of the people of a region. Article 8 of the Convention on Biological Diversity urges us to "... respect, preserve and maintain knowledge, innovations and practices of indigenous and local communities embodying traditional life styles relevant for the conservation and sustainable use of biological diversity ..." (UN, 1992).

As mentioned earlier, since the word 'tradition' in TEK often generates much debates and controversies, TEK is generally viewed as antithetical to 'scientific', which is based on rationality and universality. However, it is built around the subjective realities of cultural practices of a community, related to the larger knowledge system of their lifeworld, land, natural resources and finally environment itself. The invisible danger in the conception of tradition forces one to replace the word with indigenous. In this context, TEK is often described as irrational, obsolete, primitive, unscientific and subjective bias. Since, it is based on cultural traditions, oral and norms and values doubts rose about its significance.

Conception of 'Sacred' Space

The intersection between culture and environment in indigenous society produces a conception of 'sacred' space. Here 'sacred' is culturally defined and is related to religious myths and socio-cultural belief systems in contrast to the conception of 'sacred' often defined by religion. The conception of 'sacred', as often defined by religion, has implication on material life and material survival. The sacred space is the area where religious rituals are performed and often seen as symbolic of nature-human interconnections. In most of the indigenous communities, there are strong cultural practices of protecting forest often called the 'sacred groves' manifesting the linkage between human beings and their nature. The community's commitment to protect the sacred space lies in its commitment to conserve the sacred groves. The relationship between environment and human beings is incongruent with the traditional lifestyle of the indigenous communities. The conception of sacred groves, in fact, is the manifestation of the spiritual and cultural ethos of local communities. Most of the sacred groves have social legitimacy ensuring a balance between traditions and environment. For instance, preventing entry into the forest without permission from the traditional institutions. Strict community rules ensure that no human interference is allowed in the sacred groves. All these narrate the role of local communities in conservation of nature through spiritual beliefs and spiritual relationship between human beings and physical environment.

In the conception of sacred grove, man is located in the larger context of society. Sacred established a synergy between living and non-living things, connecting one's self with non-human entity. The construction of sacred in knowledge production is not only for environmental sustainability, but also for social sustainability. It integrates human self with ecological self, which is essential for the survival of the humanity. According to Neisser, "The ecological self is the self as perceived with respect to the physical environment: 'I' am the person here in this place, engaged in this particular activity" (Neisser, 1988: 36). Sacred groves are often the product of the animistic religious belief, however, continued after the society entered into modernity. Individual self can be developed in the larger context of the relationship with environment. The TEK is a situated or embedded knowledge where the individual self is merged with culture and environment in which individual survives. The ecological self derived from the moral persuasion of an indigenous community sees humanity's dependence on the ecology for its survival. Individual self finds its maturity and fruitfulness when it is related to the larger context of the environmental self. Individual is taken away from the self, defined by the cannons of religion by linking individual self with ecology through the conservation of sacred space. Such conception of self is antithetical to the Western notion of an individual as self-contained, autonomous and utilitarian that need not depend on others or the environment. Thus, by creating an alienated self, capitalist development take the individual self away from ecological self.

Importance of Morality

The acquisitiveness of human beings towards nature is restricted by morality and spirituality. However, the obligation to morality is due to the fear of punitive action of the nature or the local deity that resides in the sacred forest. Instead of believing in transcendental God or local deities, it is the commitment to morality and spirituality which is blocking the individualist and acquisitive tendencies of human beings in the tribal society. Such morality is derived from the primitiveness of the community. Although class formation took place in course of time, and acquisitive tendencies emerged, the tribal communities still maintain their symbiotic relationship with the nature. Ecology is not only science, but also is based on morality and ethics connecting human beings with cosmology and interrelationship between the animate and inanimate beings. Such moral commitment has produced the emergence of a new form of obligation under traditional authority where tradition play vital role. However, the intrusion of morality gives a new dimension to obligation. The preservation of sacred groves is based on the moral, physical and political obligations of the people in the tribal society.

Moral obligation is based on the moral principles to obey ecology. The cultural lineages based on environmental ethics strengthen moral obligations. Physical obligation is based on fear of punitive action by the local deity. The political obligation is based on the fear of action by the traditional political institutions.

Communitarian Ecological Self

The interaction between nature and environment produces communitarian ecological self, where the conservation of environment is based on the relational selves of the tribal communities, which are more communitarian than individualist. The communitarian self of the tribal community is merged with ecological self. In fact, the communitarian values enrich the eco-centric values. Thus, the situated self of the community produces ecological self, which often contradict the individualist anthropocentric nature of the self. The community-with-nature approach leads to a kind of inter-subjective relationship with the natural world. It is the critique of the separate self. The conservation of the ecology is the part of the cultural ethos of the community. It is the realization that the interaction between human beings with non-human beings is the foundation for the preservation of the ecosystem. The ecological self is related to the communal self as it is related to the community. The ecological self is the embedded self of the communities. Here, the self-realization is possible through the conservation of non-human entities like flora and fauna etc. According to Neisser, "the ecological self is the self as perceived with respect to the physical environment: 'I' am the person here in this place, engaged in this particular activity" (Neisser, 1988: 36). The concept of ecological self is proposed by Arne Naess, the Norwegian philosopher, used within the field of environmental philosophy called 'deep ecology' (Naess, 1985). Deep ecology conceives that humans are deeply connected with nature and they take care of the natural world.

The Western understanding of self derived from the Enlightenment is considered as individualist, self-contained, rational, and does not depend on others for their survival. It is more anthropocentric and does not consider non-human elements for the self-development. Since, it is based on hedonist utilitarianism, ecosystems, natural resources do not count. In the ecological self, an individual is identified with the surroundings – ecological self as perceived with respect to the physical environment. An individual or community develops connections with the natural environment. Both the individual and community developments depend on the cordial relationship with the nature. The individual self is embedded in community self, which is directly linked to ecological self. The sustainability of ecological self-identity brings the survival of the community, and thereby the individual.

The tribal communitarian self is linked to the planetarian self. The planetarian existence defines one's self, whether, it is individual or communitarian. Planetarian responsibility means that people behave responsibility by considering the effect of their actions on others. People are capable of deriving pleasure and meaning from contributing to causes and objectives whose scope exceeds their own personal lives (Heller, 1982: 35). They can find satisfaction and reward in helping others, particularly with the knowledge that they do so with a just and democratic system.

Everyday Life and Local Rationality

In contrast to universal rationality, it is believed that local context produces knowledge. Accordingly, the socio-cultural setting is essential for the survival of a community. The tribal worldview has created respect for animals, forests, rocks, mountains and rivers. Nehru had high appreciation for tribal culture and strived for its preservation. While he wanted to protect the distinct tribal culture, he also desired to promote the tribal's material well-being. Nehru wanted to strictly regulate the entry of the outsiders. In his five principles of tribal development, he adopted a policy of moderation. He was not ready for forceful assimilation and integration as he was apprehensive that such steps would destroy the distinct cultural values of tribal society. In this way, Nehru's approach to tribal development was rather pragmatic between isolation and integration. He stood for the protection of tribal rights to land and forests.

Traditional Knowledge and Scientific Knowledge

There are two views regarding the relationship between TEK and Western Science. The first view says that TEK and Western science are compatible as it is based on observation and establishes similarity and commonality. TEK is the accumulated knowledge based on observation, inference, and verification. However, since, it is related to indigenous communities, errors may occur and hundred per cent accuracy cannot be established. The interaction of indigenous community with physical environment in which they live has produced an experimental knowledge of its own called the TEK. Menzie and Butler identified four main characteristics of TEK that include, local, holistic, embedded, moral and spiritual. According to them, TEK is local as it is based on the specific and detailed information about areas of traditional resource use. It is holistic in the sense that all elements are viewed as interconnected and cannot be understood in isolation. TEK is embedded as it is part of a particular cultural context. It is specific not only to an ecosystem, but also to a way of understanding the world in the moral and spiritual characteristic as it is grounded in a spiritual and reciprocal relationship between the people and their environment. According to Menzie and Butler, "the natural world is often understood as sentiment and proactive and infused with spirit. . . . Practices are governed by not just a principle of sustainability for survival's sake, but by a moral sanction against waste or greed" (Menzie and Butler, 2006: 10).

However, in contrast to scientific knowledge, TEK faces certain constraints as it has the problem of recording and documenting knowledge. Further, their differentiations, different version to different communities, often make it difficult for establishing its scientific validity. As mentioned earlier, TEK is viewed as the specific knowledge to a specific cultural context. The Western scientific knowledge is derived from the positivist tradition. The positivist paradigm asserts that the real events can be observed empirically and explained with logical analysis. The positivist knowledge emphasizes on value-neutral knowledge. TEK is based on values, ethics and life practices of the indigenous groups.

Although Traditional Knowledge (TK) is an empirical knowledge based on life experiences of human beings with nature, it is often presented out of the sphere of scientific knowledge as it is based on beliefs, traditions and values. However, other prevalent view is that traditional knowledge is complementary to Western science, not a replacement for it. Noted anthropologist, Levi-Strauss viewed that there are two ways of acquiring knowledge about the universe and nature. The first is the concrete approach of Traditional Knowledge and the second is the abstract approach of Western scientific knowledge (Levi-Strauss, 1962: 269). In his opinion, "the physical world is approached from opposite ends in two cases: one is supremely concrete (indigenous knowledge) and the other supremely abstract (modern science)" (Levi-Strauss, 1962).

Feyerabend, an Australian philosopher, traces two traditions of knowledge – Abstract and Historical. While the abstract tradition is based on scientific and Western knowledge, the historical tradition is based on knowledge possessed by the people as per the cultural practices of everyday life (Feyerabend, 1987). Feyerabend was critical of Western knowledge's search for ultimate truth. According to him, knowledge makes sense only in cultural context. In other words, cultural context enables production of knowledge. Knowledge thus has certain subjective values and it is not based completely on pure reason. In broader terms, knowledge is reflection of the material world and human interaction with the nature. Western knowledge is often described as scientific knowledge as it is grounded in rationality. The Western scientific knowledge based on analytical and reductionist methods adopts positivist rigour in understanding natural phenomenon.

The ideas of both Levi-Strauss and Feyerabend clear the path towards approaching the idea of TEK. The production of culture is an intervening variable between human organisms and environment (Koreber and Kluckhohn, 1952: 186). Myths, beliefs, ethic and folkways in cultural spheres strengthen the relationship with ecology and nature. Cultural ecosystems are created by the relationship between communities and groups. Both TEK and scientific knowledge are based on observation of environment and empirical knowledge. TEK is a form of knowledge which is empirically derived from nature. TEK is holistic knowledge in which all elements of nature are viewed as interconnected and not in isolation. In this case, human beings are inseparable from nature. Since nature is inseparable from livelihood, there is a cultural component of knowledge. TEK is integral to the cultural values, spiritual beliefs and customary laws of the indigenous communities in a society. It is also situated knowledge, embedded knowledge and observational knowledge. Its rationale comes from the ethno-science in contrast to modern science. TEK is promoted to bring environmental sustainability based on the symbiotic relationship with nature.

There are fine distinctions between TEK and scientific knowledge (Table 18.1). While TEK is based on belief system, culture and lived experience, scientific knowledge is based on rationality and logic. TEK is based on spatial specific and community specific, as it is related to the larger context of time and space. In fact, identity and culture influence its production. The tradition knowledge is based on observation like the Western science. TEK is not at all related to rationalism, but often based on beliefs, customs, ethnic and related to culture. Production of TEK is often based on the subjective understanding of knowledge rather than the objective understanding of knowledge. Such a process links knowledge production as a social activity rather than a commodity production process.

Table 18.1: Difference between Traditional Ecological Knowledge and Scientific Knowledge

Traditional Ecological Knowledge	*Scientific Knowledge*
Qualitative	Quantitative
Intuitive	Rational
Holistic	Reductionist
Fusion of mind and matter	Separation of mind and matter
Moral	Value-free
Spiritual	Mechanistic
Empirical observation and accumulation of facts by trial and error	Experimentation and systematic, deliberate accumulation of fact
Data generated by resource users themselves	Specialized cadre of researchers
Diachronic data (i.e. long time series on information on one locality)	Synchronic data (short-time series over a larger area)

Source: Berkes, F. (1993), p. 4.

Both TEK and scientific knowledge are based on observations and generalizations. While TEK is located at a particular space and time as it is geographically located, scientific knowledge has universal applicability. TEK is produced in a particular place and at a particular context. Thus, territoriality and cultural specificity contribute to such knowledge. TEK is based on moral, ethical as against the rational and universal ethos of modern scientific knowledge. According to Berkes TEK encompasses: firstly, the factual knowledge about ecological components and process; secondly, knowledge put into practices about environmental issues; and thirdly, the cultural values and philosophies that define human relationship with their surrounding environment and in a wider prospect to natural world (Berkes, 2000). The indigenous communities and empirical knowledge have a close experience with nature and natural resources.

Conclusion

The advent of modernity brought dramatic changes in the relationship between human beings and environment. Anthropocentric development created imbalances in the relationship between nature and society, as the development model it puts forward is based on the exploitation of nature for the greed of the people. The reciprocal interaction between the biophysical world and the human social world got disturbed. The general perception is that traditional values retard the progress created by modernity. In other words, to be modern one has to reject tradition, beliefs, norms and morality. The onslaught of modernity on environment created an atmosphere of the denial of TEK as irrational and unscientific. Such a disturbing trend poses a grave threat to the sustainability of both nature and human beings.

REFERENCES

Berkes, F. (1999). *Sacred Ecology: Traditional Ecological Knowledge and Resource Management System*, London: Taylor and Francis.

Berkes, F. (1993). "Traditional Ecological Knowledge in Perspective", in Julian T. Inglis (ed.), *Traditional Ecological Knowledge: Concepts and Cases*, Ottawa: International Programme on Traditional Ecological Knowledge.

Berkes, F., J. Colding and C. Folke (2000). "Rediscovery of Traditional Ecological Knowledge as Adaptive Management", *Ecological Applications*, Vol. 10, No. 5, pp. 1251-1262.

EIP (1996). *Traditional Ecological Knowledge Bibliography*, Environmental Information Partnership (EIP), Moose River Basin: Ontario Ministry of Natural Resources.

Feyerabend, Paul (1987). *Farewell to Reason*, London, New York: Verso.

Heller, Agnes (1982). *Theory of History*, London: Routledge & Kegan Paul Books.

Hobsbawm, Eric (1983). "Introduction: Inventing Traditions", in Eric Hobsbawm and Terence Ranger (eds.), *The Invention of Tradition*, Cambridge: Cambridge University Press, pp. 1-15.

Koreber, A.L. and C. Kluckhohn (1952). "Culture: A Critical Review of Concepts and Definitions", *Papers of the Peabody Museum of American Archaeology and Ethnology*, Vol. 47, Harvard University Press.

Levi-Strauss, Claude (1962). *The Savage Mind*, Chicago: University of Chicago Press.

Menzie, Charles and Caroline Butler (2006). "Understanding Ecological Knowledge", in Charles R. Menzies (ed.), *Traditional Ecological Knowledge and Natural Resource Management*, Lincoln and London: University of Nebraska Press, pp. vii-20.

Naess, A. (1985). "Identification as a Source of Deep Ecological Attitudes", in M. Tobias (ed.), *Deep Ecology*, San Diego: Avant Books, pp .256-270.

Neisser, Ulric (1988). "Five Kinds of Self-knowledge", *Philosophical Psychology*, Vol. I, pp. 35-59.

Oakeshott, Michael (1962). *Rationalism in Politics and Other Essays*, London: Liberty Fund Inc.

Shils, Edward (1981). *Tradition*, London: Faber.

UN (1992). *Convention on Biological Diversity (With Annexes)*, No. 30619, Rio de Janeiro, Brazil: UN.

19

People's Perception on Impacts of Fire and Grazing on the Grassland Ecosystem of Cherrapunjee

U. Shilla and *B.K. Tiwari*

Introduction

Grassland is a landscape unit dominated by grasses (Yadava and Singh, 1986), which belong to one of the largest vascular plant families the Poaceae (Graminae) which constitutes over 10,000 species containing between 650 and 785 genera (Renvoize and Clayton, 1992; Watson and Dallwitz, 1992). Grassland ecosystems have both economic and environmental functions. The most important function of the grassland ecosystems include providing vegetation cover for the development and stabilization of soil, protecting the soil, sustain diverse vertebrate and invertebrate communities by ensuring production of feed for animals, firewood and timber plus other indirect benefits to the land users. Grassland ecosystems are mostly composed of perennial grasses and have underground storage structures that aid in survival during adverse environmental conditions (van Wyk, 2007). The above ground structures are however determined by external factors such as fire, grazing, and the amount of rainfall and availability of nutrients in the soil.

Most grasslands in humid and sub-humid regions have developed as the result of deforestation and as a result of grazing and burning, where they form sub-climax vegetation. Grasslands are maintained within a complex disturbance regime that includes fire, grazing and periodic drought. These disturbances prevent encroachment of grasslands by woody species. Either of these or a combination of all these factors is responsible for maintaining the grasslands across the globe. The biodiversity of grasslands is usually enhanced by moderate grazing and periodic fire. Fire and/or grazing plays an important role in determining the composition and structure of vegetation in the grassland ecosystems (Shilla and Tiwari, 2015). Grazers have the tendency and preference to graze on the burnt sites. Fire and grazing are known as one of the key drivers influencing structure and function of grassland ecosystems but the response of soil and vegetation communities towards fire frequency and grazing intensity varies remarkably between sites with different environmental conditions.

Study Area and Methodology

The survey was conducted at Cherrapunjee town which is situated about 50 km south of Shillong in East Khasi Hills District of Meghalaya. Cherrapunjee receives extremely high rainfall that results into loss of sediment and nutrients that impoverishes the soil.

The topography of Cherrapunjee is undulating with the land surface bearing thin topsoil and exposed rocks. Except for few patches of forests occurring in the low lying areas such as along the streams and valleys, the vegetation of Cherrapunjee mainly consists of grasses and herbaceous forbs. The grassland ecosystem is under constant anthropogenic pressure. Human interference such as annual fire, grazing, mining of coal, limestone quarry, sand extraction and other developmental activities have further accelerated the pace of the degradation of the landscape of Cherrapunjee.

The specific objectives of the study are to understand peoples' perception with regard to the extent of fire and grazing in grassland ecosystem; to substantiate the negative impacts of fire and grazing and to develop strategies for prevention of further degradation and eco-restoration of the grassland ecosystems of Cherrapunjee. In order to achieve these objectives, questionnaires were prepared. The semi-structured questionnaire for household survey was distributed to the residents of the study area and data through these questionnaires were collected at the household level. Also, interviews were conducted with the key informants with the help of a village questionnaire to obtain data on the village and the study area.

Result

Profile of Respondents

The profile of respondents covered during the study is mentioned in Table 19.1. It indicates that occupation-wise 36 per cent respondents belonged to pensioner/self-employed category, followed by respondents having government services and cattle rearing, respectively sharing 32 and 32 per cent distribution in the total number of respondents covered. Age class-wise maximum respondents belonged to the age group category of 40 to 49, followed by 60 and above, 30 to 39 and 50 to 59 category, respectively having 26, 21 and 21 per cent respondents. Literacy-wise maximum respondents belonged to high school category, followed by graduate and above, intermediate and illiterate, respectively having 26, 21 and 11 per cent respondents. Gender-wise percentage of male and female respondents covered in this study was 58 and 42 per cent respectively.

Table 19.1: Profile of Respondents

Criteria	*Category*			
Gender	*Male*	*Female*		
Respondents (%)	58	42		
Marital Status	*Married*	*Unmarried*		
Respondents (%)	89	11		
Age class	*30 to 39*	*40 to 49*	*50 to 59*	*60 & above*
Respondents (%)	21	32	21	26
Educational Status	*Illiterate*	*High School*	*Intermediate*	*Graduate & above*
Respondents (%)	11	42	21	26
Occupation	*Pensioner/Self-employed*	*Govt. Services*	*Cattle Rearing*	
Respondents (%)	36	32	32	

Source: Author's study.

People's Perception with Regards to Fire

Fire has been taking place since time immemorial and this phenomenon occurs every year during the dry season (December to February), a period when all vegetations of the grassland are dried which speed up the spread of fire. One of the positive impacts of fire is that it brings regrowth of vegetation on burned patches. However, this depends on the type of species and their tolerance to fire. If the plant species is not resistant to fire particularly the one which do not have underground structures become sparse or slowly disappear thus making the area less vegetated. Reduced grass/vegetal cover may lead to accelerated soil erosion. One hundred per cent of the respondents rued that human activities, *viz.*, discarded cigarettes, young boys playing with matches accidentally or intentionally initiate the fire. People did not say that the fire is caused by the people deliberately, though they accepted that after fire the new shots of grasses come which are more palatable. They further said that no attempt is made to control or douse the fire (Table 19.2).

Table 19.2: Frequency, Type, Season and Impact of Fire

Query	Response			
Whether fire takes place	*Yes*	*No*		
Respondents (%)	100	0		
Month/Season of fire	*Spring*	*Rainy/Summer*	*Autumn*	*Winter*
Respondents (%)	0	0	0	100
Types of fire	*Prescribed*	*Wild*		
Respondents (%)	0	100		
Change in the grassland after fire	*Yes*	*No*		
Respondents (%)	100	0		
Impact on fodder growth	*Positive*	*Negative*	*Both*	
Respondents (%)	100	0	0	

Source: Author's study.

People's Perception with Regards to Grazing

The responses of people with regards to livestock are presented in Table 19.2. The types of livestock reared was ascertained and it was found that majority (58%) of the respondents were not engaged in cattle rearing. Twenty per cent were domesticating both cow and goat and 4 per cent were only cow. With regards to the number of livestock reared, 38 per cent of the respondents engaged in this practice rear 11-20 cows, 25 per cent rear 21-30 and 25 per cent rear 31-40. Twelve per cent of respondents reared 41-50 numbers of cows. Maximum number (50%) of respondents domesticate 1-20 goats, 25 per cent 41-60 and remaining 25 per cent rear 61-80 goats each. In mode of feeding, 100 per cent of the respondents engaged in cattle rearing responded that both stall feeding and grazing were used to feed their livestock. With regards to the season for stall feeding, 78 per cent of the respondents stall feed their livestock only during rainy season and 22 per cent stall feed for whole year. Other than twigs and branches of vegetation derived from forests, grassland etc., other types of fodder, *viz.*, hay, straw, legumes, stovers (fodder) and other agricultural by-products were also given to the stall fed livestock.

Table 19.3: Livestock Population and Feeding Mode

Query	Response			
Cattle rearing	*Cow*	*Goat*	*Both*	*None*
Respondents (%)	4	0	21	58
Number of Cow	*11-20*	*21-30*	*31-40*	*41-50*
Respondents (%)	38	25	25	12
Number of Goat	*1-20*	*21-40*	*41-60*	*61-80*
Respondents (%)	50	0	25	25
Mode of Feeding	*Stall Feeding*	*Grazing*	*Both*	
Respondents (%)	0	0	100	
Period of Stall Feeding	*Whole Year*	*Rainy Seasons*		
Respondents (%)	22	78		

Source: Author's study.

The responses of people with regards to grazing are presented in Table 19.3. The study highlighted that 100 per cent of the goats grazed in the grassland. The peoples' perception with regard to per cent of cows grazed varied from 81 to 100 per cent. One hundred per cent of the respondents reported that no rules were enforced by the village administration with regard to control of grazing.

Negative Impact of Fire and Grazing

Besides the positive impact of fire, like regrowth of grasses on burned patches that is very much palatable to the livestock, 100 per cent of the respondents also pointed out the negative impact of fire giving reason that most of the grass species have become sparse due to frequent fire. Frequency of cattle being taken from one place to another depends on the availability of grasses. Grazing was mostly carried out in community land which is a common property resource. General perception of the people was that the grasslands are degrading and suitable measures need to be taken for their conservation.

Table 19.4: Area Grazed by Livestock

Query	Response	
Area of grassland covered by livestock for grazing	*Whole*	*Part*
Respondents (%)	100	0
% of cow grazing in the grassland	*Part*	*All*
Respondents (%)	0	100
% of goat grazing in the grassland	*Part*	*All*
Respondents (%)	0	100

Source: Author's study.

Conclusion

Fire and overgrazing are the factors causing negative impact to the grassland ecosystem. If the intensity of grazing is high, it can prevent recovery of the grassland vegetation and soil becomes exposed leading to accelerated erosion. However, the study revealed that the practice of cattle rearing is reducing with time as livestock population is becoming less. Occupational shift from livestock rearing to self-employment and government services is the major cause of reduction in the population of livestock in the area.

Recommendations for Management of Grassland Ecosystem

Based on the study, the following specific and actionable recommendations have been made. This also includes the strategies suggested for eco-restoration of degraded grassland ecosystems.

Fire

1. District administration need to establish a Fire Prevention and Control Committee under the chairmanship of the Deputy Commissioner/Additional Deputy Commissioner for prevention and control of fire in Cherrapunjee area.
2. A Fire Prevention and Control Committee may be established in every village by the village durbar under the chairmanship

of the village Headman. The activity of the village committees may be supervised by the Syiem of Sohra (Head of the Elaka, Traditional Chief). Khasi Hills Autonomous District Council may also be kept in the loop.

3. In every village a group of volunteers may be identified and trained in the skill of fire control and dousing.
4. Every village be provided with fire fighting equipment and tools.
5. Funds may be made available for cutting of fire lines in order to prevent the spreading of fire to adjacent areas.
6. The village community doing good work in the prevention and control of fire may be rewarded with additional social development projects or extra funds for on-going projects.
7. Awareness programme on adverse impacts of fire may be organized by the district administration to educate the local inhabitants regarding the short term and long term negative effects of fire.
8. The village authority must impose fines to individual(s) who deliberately or accidentally caused fire.

Grazing

1. Rotational grazing should be adopted to allow recovery of grassland vegetations to avoid ecological impact like soil erosion and bio-diversity loss.
2. If grazing takes place on community land and overgrazing exist, the village administration needs to regulate the numbers of cow or goat allowed to graze on such lands.
3. Stall feeding should be compulsorily implemented in villages where grasslands have excessively degraded for restoration works may be initiated by declaring the lands out of bounds for grazers.
4. In case of shortage of fodder, initiative needs to be taken up to develop and improve the grasslands by revegetating the areas with fodder species. Implementation of Feed and Fodder Development programmes, can meet the requirement of livestock besides reducing pressure on the grassland ecosystem.
5. Bamboo species can be planted for restoration of degraded lands as it is an ideal species for water and soil conservation and for carbon sequestration due to its versatility and renewability in nature (Shilla and Hussain Mir, 2016).

REFERENCES

Renvoize, S.A. and Clayton, W.D. (1992). "Classification and Evolution of the Grasses", pp. 3-37. In: G.P. Chapman, (ed.), *Grass Evolution and Domestication*, Cambridge University Press, Cambridge, UK, p. 390.

Shifting Cultivation: Sustainability and Resource Conservation in Asia, IFAD, IDRC, CIISAD, ICRAF, IIRR June, 2001.

Shilla, U. (2015). Impact of Fire and Grazing on Structure and Function of Grassland Ecosystem of Cherrapunjee, Ph.D. Thesis, North-Eastern Hill University, Shillong.

Shilla, U. and Hussain Mir, A. (2014). "Potential of Bamboo Species in Ecological Restoration of the Degraded Lands in Meghalaya, North-East India", *Biodiversity and Environmental Conservation*, Discovery Publishing House Pvt. Ltd., New Delhi.

Shilla, U. and Tiwari, B.K. (2015). "Impact of Fire and Grazing on Plant Diversity of Grassland Ecosystem of Cherrapunjee", *Keanean Journal of Science*, Vol. 4: 67-78.

van Wyk, B. (2007). *Grassland: The Most Threatened Biome in South Africa* [Online]. Available from: *www.sawac.co.za/articles/grassland.htm-10k* [Accessed: 25/06/07].

Watson, L. and Dallwitz, M.J. (1992). *The Grass Genera of the World*, CAB International, Wallingford, UK. p. 1038.

Yadava, P.S. and Singh, J.S. (1986). "Progress in Ecology", Vol. 2. Grassland Vegetation: Today and Tomorrow, New Delhi, p. 132.

APPENDIX 1

A. Village Questionnaire

1. Village Name:
   Kyrteng jong ka Shnong:
2. Location:
   Jaka
3. Total Geographical Area of the village:
   Ka jingheh jingiar jong ka shnong
4. Name of Respondent:
   Kyrteng:
5. Designation (if any):
   Ka Kam ba phi trei:
6. Age:
   Rta:
7. Education:
   Pule puthi:
8. Marital status:
   Shong kurim ne/em:
9. Total no. of households in the village
   Numbor jong Ka Jingdon jong Ki Longing ha Ka Shnong:
10. Total no. of Households (HH) rearing goat:
    Numbor jong Ka Jingdon jong Ki Longing ki ba ri jingri (blang) ha Ka Shnong:
11. Approximate number of goats in the village:
    Katno tylli eiei ki don ki Blang hapoh shnong:
12. What percentage of goats grazes in the grasslands?
    Na ka shispah katno tylli eiei Ki blang ki ba leit bam phlang?
13. Total number of Households (HH) rearing Cows:
    Numbor jong Ka Jingdon jong Ki Longing ki ba ri jingri (masi) ha Ka Shnong:
14. Approximate total number of cows in the village:
    Katno tylli eiei ki don ki masi hapoh shnong:
15. What percentage of cows grazes in the grasslands?
    Na ka shispah katno tylli eiei Ki cows ki ba leit bam phlang?
16. Any rules enforced by the village pertaining to grazing?
    Hato don ne em ki Ain ne kanun ba la shna ne pyrkhing da ka shnong halor ka jing pynbam phlang ia ki blang bad ki masi?
17. Whether fire takes place: Yes/No. If Yes, when and which month?
    Hato ka jingpluh ding ka ju jia manla u snem Hooid/Em. Lada haoid, mynno bad ha u bnai aiu.

18. History of fire (when it started)?
    Histori jong ka ding, naduh ka por jong ki mynnor ka jingthang ka sdang naduh ka por mynno?
19. Any pattern of rotation of grazing land?
    Phi ia lam bam ia ki masi bad ia ki blang haka juh beit ka jaka lane phi ialam naka wei ka jaka sha ka wei pat ?
20. Do you see any changes in the grassland after fire?
    Phi ju iohi eiei, kum ki jingkylla ha ki jingthung ynda da lah dep ka jingpluh ding lane thang?
21. In your opinion, why fire takes place?
    Ha ka pyrkhat jong phi shimet, balei ka jingpluh ding ka jia?
22. In your opinion, is it beneficial or harmful for the people of the village?
    Ha ka pyrkhat jong phi shimet, hato ka ktah ne em ia ka ioh ka kot lane hato ka don jingma ne em ia ki briew hapoh shnong?
23. What need to be done in order to stop annual fire?
    Ka lad ka lynti aiu ka ba ngi lah ban leh ban khang lane sangeh noh ia ka rukom thang ding man la u snem?

B. Household Questionnaire

Village Name:	Location:
Kyrteng jong ka Shnong:	Jaka:
Name of Respondant:	Age:
Kyrteng jong u/ka nong jubab:	Rta:
Education:	Marital status:
Pule puthi:	Shong kurim ne/em:

1. Do you rear cattle like Cow/Goat, if yes how many cows and how many goats?
   Phi ju ri jingri ne em, nuksa kum ki Masi bad ki blang, lada haoid te katno tylli ki masi bad katno tylli ki blang?
2. What is the purpose of rearing the cattle?
   Kaei ka jingmyntoi ka ba phi ioh haba ri jingri?
3. How much area do your cattle cover while Grazing?
   Katno eiei tharai phi lah ban shu batai ia ka jingheh lane ka jingiar jong ki jaka ba ki blang bad ki masi ki ju iaid leit bam?
4. Do you provide Stall feeding during spring, summer, autumn and winter or throughout the year?
   Phi ju ai jingbam ia ki jingri hapoh sem tang ha ka por Pyrem ne Lyiur ne Synrai ne por Tlang lane baroh shisnem?

5. Wherefrom do you collect the fodder?
   Phi ioh bad wan rah na kino ki jaka ia ka jing bam jong ki jingri jongphi?
6. Apart from grass do you feed them any other feeds?
   Lait na u phlang phi ju ai da u skum ne da kiwei pat ki jingbam?
7. Name the plant/grass collected for stall feeding?
   Jer ia ki kyrteng jong ki phlang ba ki jingri ki smat ban bam hapoh sem?
8. Do fire takes place every year in the grassland?
   Hato ka jingpluh ding ka ju jia manla u snem ha ki jaka lum lane madan phlang?
9. When do fire occur, which season of the year, from which month to which month of the year and at what time of the day is usually fire induced?
   Kum ha ki por aiu ka jing pluh ding ka ju jia, ha ka aiom kaba kumno ba ka jingthang ka sdang, haduh u bnai aiu haduh u bnai aiu bad katno baje eiei ka jing thang ding ka sdang?
10. Who starts the fire?
    Mano ba sdang ban ai ding?
11. How does fire start?
    Kumno ka jingpluh ka sdang hawto phi lah ban shu batai shi kyntein ar kyntein?
12. Do you take permission for inducing fire, if Yes, who gave you the permission?
    Hawto phi da ioh jingbit ne em ban thang lane ai ding, lada hooid te mano ba ai jingbit ban thang ding?
13. Any history of fire? (Since how long the grassland are being put to fire?)
    Naduh ka por jong ki mynnor lah ju don ka khanna halor kum kine ki jingthang ne kumno?
14. Do you see any change in the grassland after fire took place?
    Phi ju iohi eiei, kum ki jingkylla ha ki jingthung ynda da lah dep ka jingpluh ding lane thang?
15. Is Fire good for cattle grazing?
    Ka jing thang ding ka bha ne em ia ka jingbam jong ki masi/blang.
16. How much fodder each Cow eat when they are stall fed (full grown and young ones separately)
    Katno eiei ki masi ki bam hapoh sem (Ki masi ba lah heh bad ki masi ba dang rit la la ka jong)
17. How much fodder each Goat eat when they are stall fed (full grown and young ones separately)
    Katno eiei ki masi ki bam hapoh sem (Ki masi ba lah heh bad ki masi ba dang rit la la ka jong.

20

Perspective on Water Security Status in Manipur: A Comprehensive Review Study

Wazir Alam

Introduction

Water security is a relative and a dynamic concept that depends on both natural and anthropological interference of hydrological system and its processes (Sadoff and Muller, 2009). The concept of water security has greatest relevance to hydrological, political, cultural and socio-economic components of a watershed that have profound influences on demographic changes of a region (UN Water, 2015). The freshwater availability in many parts of the world is a major concern as it contributes only 2.5 per cent of global water balance and support as a major source of drinking and irrigational water. Out of 2.5 per cent, only 0.4 per cent contributes as surface water and 0.31 per cent contributes as groundwater (of which only 0.3 per cent is economically extractable) (Todd, 2004). Moreover, spatio-temporal accessibility, availability and quality of water also further ensure water security of a region in any part of the world (GWP, 2012). There is no comprehensive solution to the world's water security crisis, as various factors that may originate the challenges in the form of food security, demographic changes and migration, urbanization, industrialization, climate change; overexploitation or indiscriminate use of water resources, inadequate management practices, lack of people's awareness and public participation (UN Water, 2015). However, the solution to the challenges of water security can be achieved at local, regional and global level through combined effort of technical innovations, investment, policy shift and participation of various stakeholders (Holmgren, 2015). The water security of North-Eastern Indian states is highly prone to the consequences geo-ecological fragility, regional climate change, strategic location of Himalayan landscape, trans-boundary river basins, regional and trans-boarder water sharing and other socio-economic instabilities (Baruah *et al.*, 2003; Das *et al.*, 2011). The vulnerability of hydrological regime of the Brahmaputra and Barak (Meghna) river systems (both source and resource) are critical (INCCA, 2010). It is expected that by 2030, the water vulnerability index of Manipur such as floods frequency, flash and urban floods, landslides are expected to increase by 25 per cent to the present situation (Das, 2011, Ravindranath *et al.*, 2011). The forest cover areas of the state already decreased from 79.21 per cent to 76.10 per cent during 1987 to 2013 (FSI, 2015), that can have profound influence on hydrometeorological processes and local climatic of the state. The monsoons contribute around 59 per cent of total rainfall in the region that feeds the water bodies and constitute as a major water supporting system to the people of the state. So, it is expected that any form of changes in rainfall pattern or other climatic factor can initiate the water security problems in the region in the form of scarcity of drinking and agricultural water, crop failure, flash flood, urban floods, health security problems etc. (MSAPCC, 2013).

Environmental Description of Manipur

Geographically, Manipur is located in a peculiar location of Indo-Myanmar arc system between latitudes 23.80° N to 25.68° N and longitude 93.3° E to 94.78° E and is known to be among the world's heaviest rainfall region. The region receives most of its rainfall from SW monsoon during month of June-September with annual average of about 1436 mm (accounts for 59 per cent of total rainfall in the state). Geomorphologically, the state is characterized by two distinct physical regions – a narrow valley/flat plains covering 10 per cent of total geographical area contributing 59 per cent of the population that is surrounded by the rugged hills. Physiographically, the hills that extend from NNE-SSW direction with a slope of 0° to 37°. The state experiences sub-humid and sub-tropical to temperate climate with appreciable spatial variation exhibiting a highly contrasting meteorological condition in various seasons. The annual average minimum and maximum temperature of the state ranges between 1° C to 36° C respectively with coldest temperature –3° C during December and January (MSAPCC, 2013).

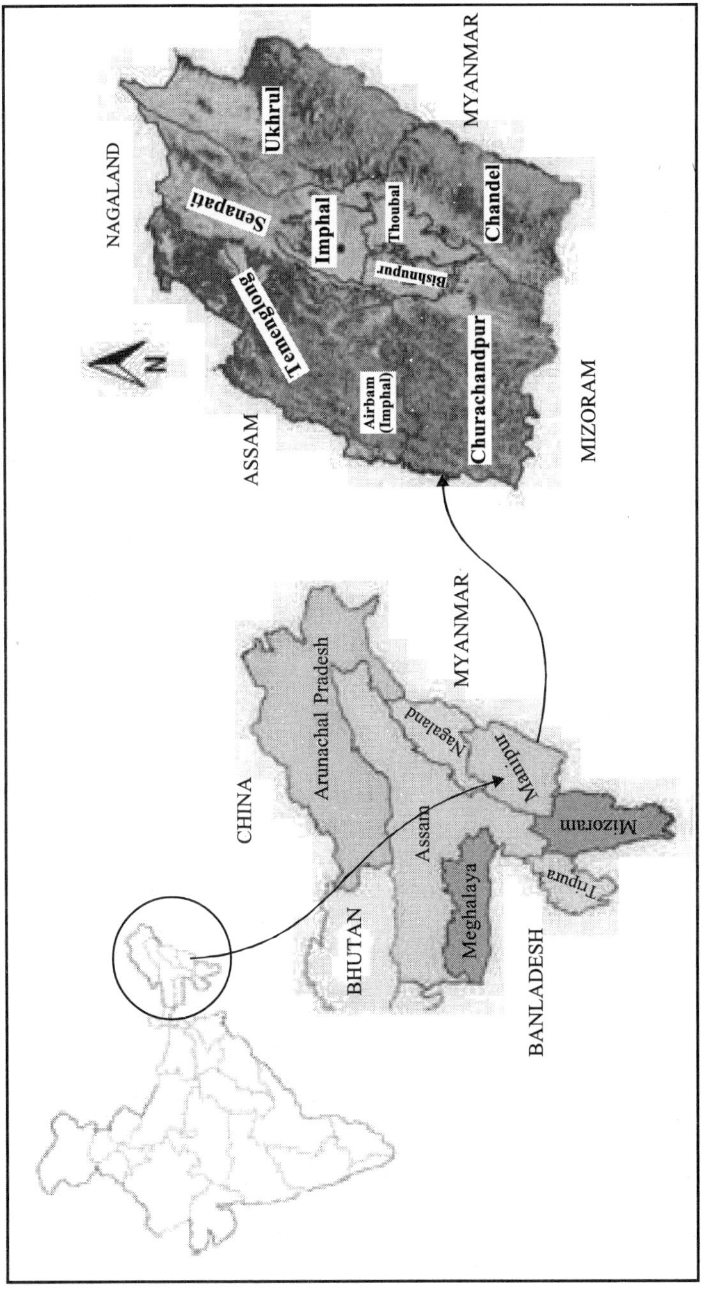

Figure 20.1: The Map Showing the Location of Manipur.

Water Resources of Manipur

The population of state is mainly dependent on surface water resources which comprises of rivers, streams, wetlands, ponds etc. These water resources supports the population residing in 1820 sq.km. of flat valley and 20,507 sq.km. hilly terrain for drinking and irrigation purposes. Presently, due to changing pattern various natural and anthropogenic factors the hydrological regimes are also altered causing water imbalance in the region. These changes are already visible in the form of accessibility and affordability to quality of drinking and agricultural water, triggering threat to water security of the state. Moreover, unplanned urbanization, unsustainable water management practices and deficit of concrete water policies have aggravated the problems of water scarcity to a great extent.

Status of Surface Water Resources

The state comprises of two major river basins, viz., the Barak River basin draining western part of the state with average water potential of 3,295 MCM per annum (covering 9,042 sq.km.) and the Manipur River basin draining the eastern half of the state including the Manipur valley with average water potential of 5,192 MCM per annum (comprises of 6,332 sq.km.).The average availability of surface water resources of two river basins and wetlands accounts for 18,487 MCM per annum (CWC, 2015). The other major rivers system and its tributaries include Thoubal, Imphal, Nambul, Makru, Leimatak, Irang, Iril, etc. The rivers Imphal, Nambul, Iril and Thoubal which flow through the Manipur valley are in grip of environmental threat due to considerable reduction in hydrological flow regime, water level, discharges, drying of sources and deterioration of water quality in recent years (PHED, 2009; CPCB, 2013; Koijam & Yumnam, 2015; ENVIS, 2015; Alam and Laishram, 2017). Moreover, anthropogenic disturbances in the catchment area such as quarrying, deforestation and destruction of forest have resulted in heavy siltation of these river systems leading to gradual disappearance streams and other water bodies causing heavy run-off in the form of flash and urban floods (INCCA, 2010). Presently, Government of Manipur is planning for nearly twenty mega dam under Manipur Hydroelectric Power Policy, 2012 including proposed controversial 1500 MW Tipaimukh multipurpose hydroelectric power project which is projected to submerge 27,000 hectares of forest land including felling of more than eight million tress (Koijam & Yumnam, 2015). Wetland (locally known as "pats") of Manipur covers nearly about 2.85 per cent of total geographical area and plays a very crucial role in the socio-economic and cultural aspects of life of the people. Altogether 708 wetlands with geographical coverage of 63,616 ha are recorded in the state. Out of total wetlands, 167 wetlands are of an area greater than 2.25 ha and 541 are of smaller than 2.25 ha. These wetlands cover 61.5 per cent area under ponds, lakes; 26.2 per cent under streams and river plain and 5.5 per cent area under waterlogged and aquaculture pond. Presently, Directorate of Environment, Government of Manipur recognized and identified 19 major wetlands for conservation (MSAPCC, 2013).

Table 20.1: Status of Major River System of Manipur

River Basin	*Major Rivers*	*Catchment Area (km²)*	*Percentage of Average Yield*	*Yield in MLD**
Barak River Basin	Barak	6,865	43.0%	230
	Tuvai River	1,860	17.0%	95
	Jiri River	316	7.3%	39
	Total	**9,041**		**364**
Manipur River Basin	Manipur River (up to Ithai Barrage)	5,109	21.0%	116
	Manipur River (beyond Ithai Barrage)	1,223	10.0%	55
	Total	**6,332**		**171 MLD**

*MLD = Million Litres Daily.
Source: ENVIS, 2015, India.

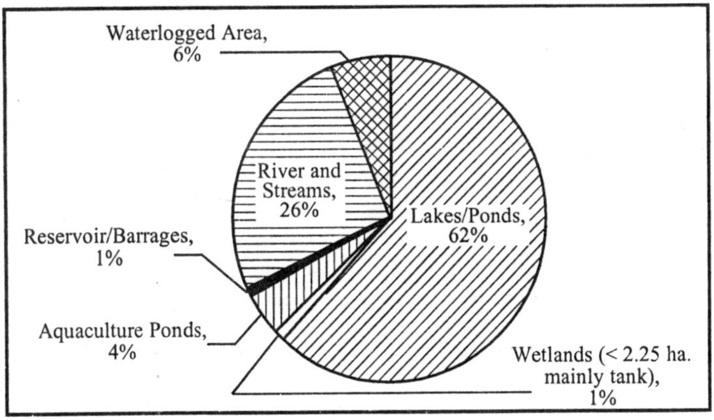

Figure 20.2: Distribution of Water Bodies of Manipur.

Source: ISRO, 2010.

Status of Groundwater Resources

Manipur have recent geological origin and comprises of flat alluvium filled valley of Fluvio-lacustrine origin and surrounded by denude-structural hills. The valley consist intermontane piedmont superficial alluvial plain underlying by tertiary rocks of Barail ranges which signifies mostly a depositional environment consisting of mainly of sandy and clayey formation under water table condition with varying depth of shallow aquifers between 12-25 m, medium aquifer 60-65 m and deep aquifer 65-75 m, respectively (CGWB, 2008). Due to clayey nature of the top aquifer in most of the valley region, groundwater extraction is not a suitable or viable option for irrigation and drinking purposes (CPCB, 2013). Recent preliminary investigations on groundwater quality of some selected pockets of Imphal East, Imphal West, Thoubal and Bishnupur districts of valley region have revealed the presence of harmful lithophilic ions such as Arsenic (As), Fluoride (F) and other trace and heavy metal in groundwater (Oinam and Ramanathan, 2012; ENVIS, 2015; Alam and Neihsial, 2017). Moreover, the various research findings have indicated that arsenic and fluoride affected habitations of Manipur valley are mainly located within the newer alluvium aquifer and geochemical composition is controlled by natural weathering process (Oinam and Ramanathan, 2012; Alam and Neihsial, 2017).

Water Security and Climate Change

The phenomenon of climate change are now apparent all-over the world and its impacts are being witnessed in every aspect of human security (Morrison *et al.*, 2009). In India, climate change is expected to impact the hydrological flow regime of Ganga-Brahmaputra-Meghna (Barak) river basin (CWC, 2015). Moreover, it is expected that climate change will influence the monsoon patterns in the region in the form of more intense and heavy rainfall, however will concentrate on fewer rainy days, increasing the extreme events, flash floods and other hydrometeorological disasters (INCCA, 2010). In spite of receiving sufficient rainfall, the state has witnessed several droughts like situation in last few decades due to lack of appropriate and adequate water management initiatives (ENVIS, 2015; Koijam & Yumnam, 2015). INCAA projected that average mean annual rainfall will vary from minimum 940 ± 149 mm to a maximum 1330 ± 174.5 mm and there will be increase of in precipitation by 0.3 to 3 per cent in whole Manipur by

2030. Thus, will reflect in the form of high intensity rainfall (i.e. 1-6 mm/day) as numbers of rainy days are likely to decrease by 1-10 days. The northern parts of state are going to experience 19 per cent increase in rainfall in next 30 years (Ravindranath *et al.*, 2011). Thus, it can be expected that uneven, untimely and high intensity rainfall will increase water induced or hydrological disasters such as floods, flash floods, landslide, scarcity of drinkable water, agriculture crop failure, water-logging and incidence of vector and water-borne diseases. The changes in rainfall pattern have already impacted run-off and siltation of river beds, flash floods, groundwater recharge and quality in alluvial aquifers (Alam and Neihsial, 2017).

Water, Agriculture and Food Security

Agriculture accounts for more than two-thirds (70%) of global freshwater use, of which the developing countries account for nearly about 90 per cent (FAO, 2012). Population growth and economic developments are significant driving factors for increasing agricultural and industrial demand for water. So, a change in water security scenario will have direct and immediate impact on agriculture and food security of any region. Many researchers and IPCC have already predicted that there will be 10-40 per cent losses in agriculture productivity in India by the end of 21st century due to human induced climate change, weather variability, unpredictability of monsoon and extreme events such as droughts, floods, cyclones etc. Moreover, it has been predicted through crop simulation model at IISc, Bengaluru, that there will be decrease in rain-fed rice production by 0.1 to 4 per cent in rice producing states of India (Ravindranath, *et al.*, 2011). In Manipur, 70 per cent population constitutes agrarian society where two types of agricultural practices are witnessed, i.e. settled and organized agricultural practices in valley districts, foothills, terrace slopes and shifting cultivation (*Jhum*) practices in hills and hill slopes. As per 2012-13 agricultural data, the state's predicted food production was around 688 thousands MT against demand of 720.17 thousands MT. As per prediction study carried out at IISc, Bengaluru, out of nine districts, seven districts of state, *viz.*, Chandel, Thoubal, Imphal West & East, Senapati, Ukhrul and Tamenglong, the rice production will decrease by 2050 due to various fluctuations and variability of hydrometeorological factors (Ravindranath, *et al.*, 2011). These factors can be in the form of late monsoon, unpredictable seasonal rainfall and drought like situations that may delay rice seeding and transplanting. Moreover, untimely rain and flood will inundate agricultural field during Panicle Initiation (PI) stage of rice causing decrease in productivity (MSAPCC, 2013).

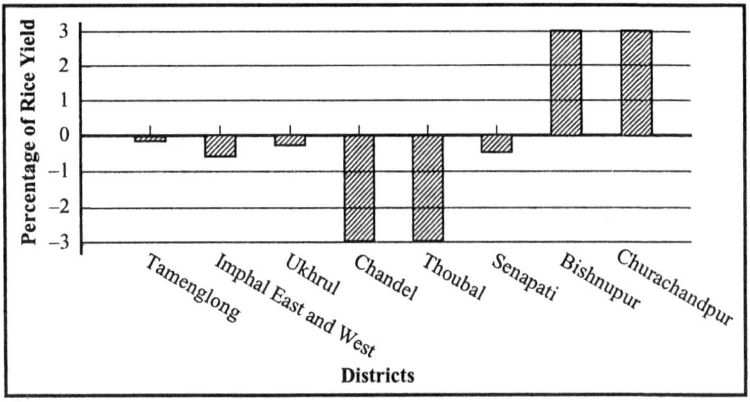

Figure 20.3: Projected Percentage of Rice Yield of Manipur.

Source: Ravindranath, *et al.*, 2011.

Manipur is having 63,616 ha of geographic area under the different types of wetlands, ponds and waterlogged areas which are sources for fishery and aquaculture development (MASTEC, 2007). But due to present threats from both natural and anthropogenic factors such as erratic rainfall, extreme climate events, heavy siltation of water bodies, nutrients and pesticides runoff from agricultural fields and pollution of feeding rivers to these wetlands have threatened the loss of habitation of fishes and aquaculture practices in the state (ENVIS, 2015). The Government of Manipur has a vision to produce 35,000 MT of fish by 2017 against the present fish requirement of 30,000 MT (MSAPCC, 2013), which will be a challenging objective, as the fate of fishery sector is completely dependent on the water security challenges. Increasing extreme weather events and inconsistent rainfall pattern, may also impact the water intensive sector like horticulture, livestock and animal husbandry in the state. However, present weak and delayed monsoons have already impacted the natural cycle of the crops, creating challenges to state's food security, agricultural productivity that could be catastrophic to the state economy and quality of life if appropriate water and food security initiatives are not taken immediately (MSAPCC, 2013). The Government of Manipur has planned to cover 43,640 ha, over existing irrigation potential of 6,550 ha of land, which not only depends on the success of implementation of various major irrigation projects such as Chakpi Multipurpose Project and Jiri Irrigation Project (with potential

22,150 ha water supply) (MSAPCC, 2013), but the major challenges will be in the form of distribution of rainfall and climatic variability in the region.

Water Security, Economy and Livelihood

Around 58 per cent of population of India belongs to an agrarian society or directly or indirectly dependent on agriculture sector and it contributes one-third of India's GDP (GWP, 2012). Undoubtedly, the threat to water security will be directly apparent itself in state's economy and livelihood. Extreme geo-environmental phenomenon related to water and hydrological regime will result in extreme weather events, floods, drought like situation, delay and erratic monsoons, cyclones and untimely high intensity precipitation event that will have negative impacts on state economy, livelihood and agricultural sector. Around 70 per cent of population in Manipur living in rural area have agriculture as their livelihood option. Therefore, challenges of water security will have direct and immediate impact on rural livelihood and agriculture of the state. Agriculture sector and natural resources from forest and wetland contribute 73.82 per cent of state's GDP. Moreover, any new venture or industrial development in state will strongly consider the availability of water for their functioning and sustenance. Altogether, challenges of water security will have an immediate and irreversible impact on the state's economy, livelihood of thousands of people and their families (ENVIS, 2015).

Water and Health Security

Due to lack of accessibility to safe drinking water and sanitation, water-borne diseases contribute to as many as 80 per cent of all disease in South Asia and most vulnerable group includes particularly the young children, whose health is affected severely and often do not fully recover from the effects of water borne-diseases (UN Water, 2015). The climate of Manipur is basically sub-humid, sub-tropical and temperate type with appreciable spatial variability that makes the region more vulnerable to water-borne and vector-borne diseases. Malaria cases in the state is in increasing trend with 32 per cent contribution in Tamenglong district, followed by Churachandpur district (22%) and Imphal East (16%) with other districts contributing 1-8 per cent respectively (ENVIS, 2015) (Fig. 20.4). Due to recent climatic variability and erratic rainfall pattern, there is a increase in transmission window (10-12 months) of malaria from baseline scenario of 1960–90 (i.e. 7–9 months).

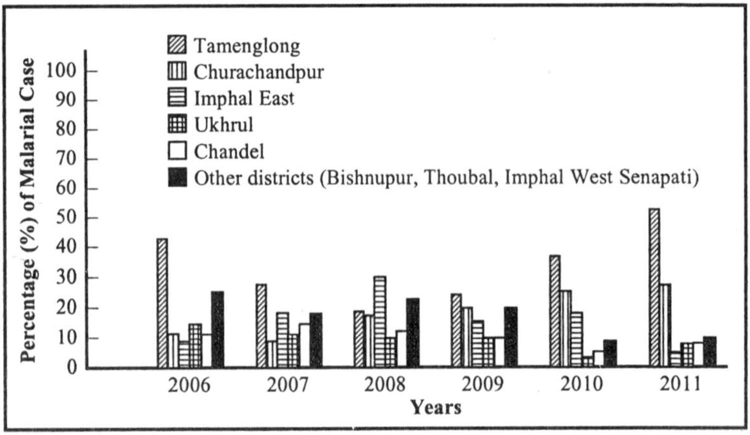

Figure 20.4: Percentage of Increase of Malarial Cases in Districts of Manipur from 2006-2011.

Source: ENVIS, 2015.

527 villages are vulnerable to vector-borne diseases that accounts for 2.6 lakhs population and are at high risk. Presently, the state is vulnerable to six types of vector and water-borne diseases (i.e. Malaria, dengue, Japanese encephalitis, chikungunya, *Kala-azar* and filariasis), which are directly related to water for its spread and transmission (MSAPCC, 2013). Moreover, the extreme weather phenomenon and hydrometeorological disasters such as excess summer rainfall, decrease in winter rainfall, intense monsoon, increases of flash floods, landslides, drought like situations and water-logging in several parts of the region are contributing for generation and transmission of these diseases by facilitating the breeding ground and their spreading. Improper drainage and unplanned urban growth also have reduced the natural flow system of water bodies increasing vulnerability of people to these types of diseases. The present trend of municipal water supply is inadequate to meet the growing demand for safe drinking water have also increased the vulnerability of people to disease and infections. The mostly affected groups include rural population, farmers, backward and weaker section of society due to lesser accessibility to safe drinking water (Kumar & Bharat, 2014). Only 40.5 per cent (1163 habitations out of 2870) are covered with drinking water supply facility that covers population around

2.5 million. Rest 42.7 per cent (1225 habitation) are partially covered and 16.8 per cent (482 habitations) are uncovered (PHED, 2009).

Water Security and Social Challenges

Manipur's index of social and economic infrastructure is 75.39 (MSAPCC, 2013). Water scarcity scenario in semi-urban and rural areas of the state is alarmingly increasing. Moreover, in last two decades more and more rural residents are forced to abandon their rural area and moving towards urban and semi-urban locations of the state due to environmental induced migration or lack of security in the agriculture sector or for other livelihood opportunities (Table 20.2). Among all these factors, water security is one of the major and initiating factors for migration of population not only within the state, but also from other parts of the country. This has already aggravated serious human security and social problem in parts of the state. The migration have already gravitate towards urban areas leading to exacerbate the existing tensions of ethnic and social divides resulting in increase of cases of strikes, blockades and crimes. Growing population will compete for limited, scarce resources and jobs leading to conflict between 'indigenous *versus* migrant', which is already prevalent and could manifest to more violent situation.

Table 20.2: Growth of Decadal Urban Population in Manipur

Years	*Number of Urban Area*	*Total Population*			*Population Percentage (%)*	
		State	*Urban*	*Rural*	*Urban*	*Rural*
1961	1	7,80,037	67,717	7,12,320	8.68	91.32
1971	8	10,72,753	1,41,492	9,31,261	13.19	86.81
1981	32	14,20,953	3,75,460	10,45,493	26.42	73.58
1991	31	18,37,149	5,05,645	13,31,504	27.52	72.48
2001	33	23,88,634	5,70,410	18,18,224	23.88	76.12
2011	33	27,21,756	8,22,132	18,99,624	30.21	69.79

Source: Census of India, 2011.

The Future

Water security is an emerging problem for the state of Manipur and it requires a holistic interdisciplinary approach to address all dimensions of water security, i.e., water, food, health and overall human security. The immediate initiatives are required to recognize the apparent problems of water security that might cause potential impacts in near future. Scarcity of water will have direct and immediate impacts on food security, agricultural security, environmental migration, human conflicts and biological diversity. A long term and sustainable collaboration through joint action are required between neighbouring states and countries for conservation of trans-border rivers, water bodies and watershed. Since, there is no comprehensive solution to the state's water security crisis and problems, the need of the hour is to formulate immediate solutions at both micro- and macro-levels simultaneously including various stakeholders such as decision and policy-makers, officials and administrators, NGOs and CBO, communities etc. In addition to that, combined effort of technical innovations, investment, policy shift and participation of various stakeholders are highly required for formulation of an amicable and sustainable solution for water security challenges for the state.

REFERENCES

Alam, W. and Neihsial, M.T. (2017). *Qualitative and Quantitative Assessment of Groundwater of Thoubal District of Manipur, India*, Proceedings of National Seminar on Science & Technology for National Development, ISCA, Imphal Chapter, 2-4th April, 2017, Manipur University, 4.

Alam, W. and Laishram, R.J. (2017). *Spatial and Temporal Assessment of Water Quality of Nambul River within Imphal City, Manipur*, Proceedings of International Conference on Natural Resources Management and Technology, 27-29th March, 2017, Centre for Advanced Study, Manipur University, 108-109.

CPCB (2013). *Status of Water Quality in India-2012*, Monitoring of Indian National Aquatic Resources Series: MINARS/36/2013-14, Central Pollution Control Board, New Delhi.

CWC (2015). *Water and Related Statistics*, Water Resources Information System Directorate, Ministry of Water Resources, Government of India, New Delhi.

Das, A.; Ghosh, P.K.; Choudhury, B.U.; Patel, D.P.; Munda, G.C.; Ngachan S.V. and Chowdhury, P. (2011). *Climate Change In North-East India: Recent Facts and Events – Worry for Agricultural Management*, ISPRS Archives XXXVIII-8/W3 Workshop Proceedings: Impact of Climate Change on Agriculture, 32-37.

FAO (2012). *Coping with Water Scarcity: An Framework for Agriculture and Food Security*, Food and Agriculture Organization of the United Nations (FAO), Rome.

FSI (2015). *Indian Statistical Forest Resources*, Forest Survey of Indian, Government of India.

ENVIS (2015). *Manipur – Status of Environment and Related Issues*, ENVIS Centre Manipur, Directorate of Environment, Ministry of Environment and Forests, Government of India. *http://www.manipurenvis, nic.in/stateprofile.*

Global Water Partnership (GWP) (2012). *Increasing Water Security – A Development Imperative*, A GWP Technical Committee Perspectives Paper, GWP, Stockholm.

Holmgren, T. (2014). *Stockholm Water Front: A Forum for Global Water Issues, March-2014*, International Water Institute, Stockholm, Sweden.

INCCA (2010). *Indian Network for Climate Change Assessment Report*, Ministry of Environment and Forests (MoEF), Government of India, New Delhi.

Koijam, P. and Yumnam, J. (2015). *Climate Crisis and Indigenous People Rights in Manipur*, Centre for Research and Advocacy, an indigenous Peoples human rights organization in Manipur, Imphal, Manipur.

Kumar, S.V. and Bharat, G. (2014). *Perspectives on a Water Resource Policy for India*, The Energy and Resources Institute (TERI), New Delhi.

MASTEC (2007). *Report on Water Harvesting and Water Conservation in Imphal East I Block, Imphal East District, Manipur*, Manipur Science, Technology and Environment Council (MASTEC), Government of Manipur.

Morrison, J.; Morikawa, M.; Murphy, M. and Schulte, P. (2009). *Water Scarcity and Climate Change: Growing Risks for Business and Investors*, A Ceres Report, Boston.

MSAPCC (2013). *Manipur State Action Plan on Climate Change*, Directorate of Environment, Government of Manipur, Imphal, Manipur, India.

Oinam, J.D.; Ramanathan, A.L. and Singh, G. (2012). "Geochemical and Statistical Evaluation of Groundwater in Imphal and Thoubal Districts of Manipur, India", *Journal of Asian Earth Sciences*, 48: 136-149.

PHED (2009). *National Rural Drinking Water Programme (NRDWP)*, Public Health Engineering Department, Government of Manipur.

Ravindranath, N.H.; Rao, S.; Sharma, N.; Nair, M.; Gopalakrishnan, R.; Rao, A.S.; Malaviya, S.; Tiwari, R.; Sagadevan, A.; Munsi, M.; Krishna, N. and Bala, G. (2011). "Climate Change Vulnerability Profiles for North-East India", *Current Science*, 101(3): 384-394.

Sadoff, C. and Muller, M. (2009). *Water Management, Water Security and Climate Change Adaptation: Early Impacts and Essential Responses*, Background Paper No. 14. Global Water Partnership, Stockholm.

Todd, D.K. (2004). *Ground Water Hydrology*, 2nd Ed., John Wiley & Sons (Asia) Pvt. Ltd., Singapore, pp. 282-302.

UN Water (2015). *Coping with Water Scarcity: Challenge of the Twenty-First Century*, UN Water Annual Report, UN-Water Technical Advisory Unit, Geneva, Switzerland.

INDEX

Acquisitive tendencies, 213
Activism, 6
Afforestation, 91, 95, 98
Agrarian society, 237, 239
Agricultural crops, 43, 108
Agricultural development, 20
Agricultural practices, 30, 108, 111, 237
Alder tree, 67
Alternative markets, 131
Ancestral property, 35, 45, 155
Ancient civilizations, distribution systems of, 26–27
Ancient knowledge, repositories of, 26–27
Angami Perception of Nature, 144
Angami way of life, 146
Anthropocentrism, 116
Apatani traditions, 30
Appeasements, 140, 147
Aquaculture development, 238
Archery, 161
Arctic ice, 188
Arts and Crafts of Nagaland, 149–50
Autonomous District Council, 51, 75, 105

Baghmara Reserve Forest, 107
Balat Border Haat, 127, 131
Bamboo basket, 161–62, 174
Bamboo poles, 109, 175
Belief systems, 25, 27–30, 146, 171, 174, 176, 211, 217
    religious, 175
    socio-cultural, 212
    traditional, 28
Betel nuts, 129–30, 132, 156, 169
Biocentric worldview, 112
Bio-diversity, 40, 61–62, 68, 74, 86, 106, 208, 211, 221
    conservation efforts, 74
Biological diversity, 211, 242
Biosphere reserves, 106, 111
Birth rituals, 173
Bombay Natural History Society, 68
Border haats, 127, 129–32
Border markets, 128–29, 131
Borok Community, 171, 178, 180
Brundtland Commission, 26, 111

Castanopsis kurzii, 80–81
Castanopsis tribuloides, 80–81
Catchment areas, 10, 30, 234
Cattle rearing, 222–25
Cause-effect, 145
Cave conservation, 188
Cave Lores, 182, 184–85, 187, 190
Caves and Environment Conservation, 188
Champhai forest range, 91–92
Chipko Movement, 43–44, 59
Civil Rights Act, 21, 24
Civil society organizations, 8, 10
Clayey nature, 236
Climate change, 3, 25, 59, 231, 236
Climatic variability, 239
Collective action, 3, 5–6
Collective identity, 5–7
Colonialism, 39
Commercial forestry, 9
Commercial timber, 13
Common Property Resources (CPRs), 44, 52, 54
Communal land, 39, 52
    quality of, 52
Communitarian, 214–15
Communities
    cultural, 207–8, 211
    ethnic, 12, 208

human, 178
non-tribal, 178
Community forest management systems, 75
Community-managed forests, 85
Community-with-nature approach, 214
Conservation
activities, 70, 72
approach, 68
effort, 63, 66, 72
efforts, 63, 6668–70, 72
initiatives, 37
issue, 72
Conserving land resource, 102
Constitutional Amendment Act, 24
Consumerism, 13
Cosmic dimensions, 145
Cosmopolitanism, 7
Cross-cultural dialogue, 144, 149
Cultural
analysis, 192
beliefs, 114
diversity, 211
ethos, 163, 212, 214
identity, 211
life, 50, 59, 171
pluralism, 17
practices, 113, 157, 212, 217
traditions, 19, 180, 212
transmission, 209–10
values, 210, 217–18
Customary
laws, 35–36, 53–54, 217
practices, 51–52, 157, 159, 180

Dams, 4, 12
Dances, traditional, 48
Data Collection and Analysis, 77
Death ceremony, 140
Deforestation, 11–12, 59, 64, 180, 189, 220, 234
Deities, 140–41, 156–57, 160, 175
Demons, 184, 188
Department of Environment, 75
District Educational Research and Training (DERT), 122
District Institute of Education and Training (DIETs), 114–17, 119–21
Disturbance index, 78–79, 86
Domestic violence, 36
Droughts, 236–37, 239–40
Dual forest policies, 51
Durkheim, 157
Dynamic mechanism, 154

Eco-criticism, 190–92, 195
Eco-critics, 192
Eco-feminism, 39, 44, 193
Ecological
conservation, 9, 207
crises, 4–5
disaster, 152
movements, 7, 9
problems, 6, 192
refugees, 10
self, 213–15
Ecology, 3, 5–6, 8–9, 25, 117, 171, 191–92, 199–200, 202, 204, 210–11, 213–14, 217
conservation of, 207, 211
cultural, 191, 207
movements, 8–9, 191
Economic
hardships, 46, 53–54
life, 59, 61, 72, 74
traditional, 60
Eco-theandric vision, 199
Eco-tourism, 68
Elaeocarpus acuminatus, 80–81
Encroachment, 81, 220
Entry Point Activity (EPA), 93–94, 96
Environment(al)
activism, 3, 6, 14
anthropogenic, 114
attitudes, 114
beliefs, 112
challenges, 3
clean, 72
conditions, 108, 220–21

consciousness, 5, 7, 14
conservation of, 59, 113, 188, 209, 214
crises, 3–4, 8, 192
cultural, 34
degradation, 3, 8–9, 12, 199, 204
depositional, 236
destruction, 10
distinctive, 198
elements, 59, 61, 158, 160
ethics, 5, 207–8
fauna, 159
functions, 220
governance, 36
greener, 67
healthy, 13–14
human, 113
impact assessment, 68
information partnership, 210
literacy, 114
local, 68
loss, 38
-man relationship, 43
management, 39
migration, 242
movements, 3, 6–13
philosophies, 180, 214
physical, 191, 212–15
policy, 38
politics, 7
pollution, 59, 191
problems, 4–5, 8–9, 12–13, 38, 113, 193
protection, 101
protest movements, 11–12
regeneration, 39
science, 202
sociologists, 4
thinkers, 39
threat, 234
toxicologists, 192
values, 114–15, 173

EPA fund, 94–96
Evil spirits, 139, 187

Ficus ecology, 199, 204
Fire Prevention and Control Committee, 225
Flash floods, 199, 232, 236–37, 240
Flora, 11, 28–29, 47, 66, 68, 71, 103, 106, 154, 158, 183, 190, 214
Flowers, 12, 156, 172–73, 183–85, 193
Fluoride, 236
Fodder, 36, 49, 54, 94, 224, 226
Folk songs, 61, 176–78
Folklore, 60, 71, 137, 156, 182–84, 194–95, 210
heritage, 183
Food security, 231, 237, 242
Forest
based industries, 12
broad-leaved humid, 76
burning, 37
categories, 81
climax, 76
Conservation Act, 23
conservation strategies, 75
degraded, 49, 91
dense, 13, 105
depleting, 13
development agencies, 91, 98
disturbance, 79
dry deciduous, 164
evergreen, 103
food-related, 37
foods, 47
gods, 28
lands, 6, 23, 51, 54, 90, 92, 97–98, 104, 234
management, 49, 51–54, 78, 90
natural, 54, 75
officials, 93, 97
open, 105
patches, 104, 221
policies, 51
private, 51, 104
products, 37, 46, 54, 76, 129
protected village/community, 105
protected, 51, 85–86, 104
protecting, 212
related activities, 44, 50

resource management, 43
restricted, 76, 78, 80, 86
rights and management, 51
rugged, 52
safeguard, 13
thick, 107
tropical rain, 12, 74
tropical semi-evergreen, 164
unclassed, 51
vegetation attributes, 86
verdant, 29
village, 76, 78
virgin, 37
world, 75
Fringe forest areas, 47
Fuel-wood, 36, 46–47, 49, 52, 75–76, 87
harvesting, 78

Gender division of labour, 37, 39–40
Gender inequalities, 40–41
Gender Relation, 34, 40
Global Environment Facility (GEF), 110
Global identity, 6
Gram Sabha, 24
Grass vegetation, 160
Grasses
broom, 36, 49
thatched, 109–10
Grasslands, 220–22, 224–26
Green Village, 68–72
Gross Domestic Product (GDP), 43

Haats, 129–30, 132–33
Haat Management Committee, 132
Happiness index, 26
Harmony, 26, 48, 147, 156, 169, 191, 194
Hedonist utilitarianism, 214
Herbs, traditional, 75
Hindu gods, 175
Horticultural, 20, 128
House construction, 52, 60, 64, 165
Human life, 48, 171
Human rights, 16–17
Human-environment interface, 34
Humanistic science, 146
Human-made environments, 114
Human-nature
equality, 121
relationship, 171, 211
Hunting, 12, 37, 47, 61–66, 69–70
Hydrological regimes, 232, 234, 239

Illegal
hunting, 65–66
logging, 85–86
Immovable property, 35–36, 138
Indigenous
belief, 146
communities, 60, 74, 109, 207, 212–13, 215, 217–18
knowledge, 169, 216
Individualist anthropocentric nature, 214
Industrialization, 3–4, 12–13, 75, 231
Infrastructural developments, 93, 95, 98
Infringement, 52
Integrative nature, 115
Interconnected nature, 199
Interdisciplinary nature, 115
Intrinsic values, 112–13, 193
Iron smelting, 129

Jhum cultivation, 19, 108, 140–42, 148, 176–77, 237
Joint Forest Management (JFM), 90–91
Joint Forest Management Committees (JFMCs), 91
Joint Forest Management, 91

Karbi
culture, 138
society, 138, 141–42
belief, 46
culture, 48, 163
ethics, 157, 162

faith, 29, 185
family, 45
folklore, 47, 164, 183–84
Hills Autonomous District Council, 226
legendary conception, 155
livelihood, 155, 158, 161–62
markets, 129
myths, 159
pine trees, 109
religion, 185, 195
rural women, 48
sacred forests, 27
tradition, 29, 161
treatment of nature, 195
villagers, 129
Khonoma, 60–64, 66–72, 200
Kitchen waste, 30
Knowledge
advanced agricultural, 29
component, 95
embedded, 213, 217
empirical, 208, 216–18
environmental, 113, 168, 210
experimental, 208, 216
human, 154, 169

Laitkor forest Range, 28
Land Acquisition Act, 23
Land ownership, 51–52
Livelihood, 4–5, 8, 10, 14, 44, 48, 53–54, 59, 61–63, 69–71, 75, 81, 108, 110, 239
Livestock population, 225
Logical analysis, 216

Male chauvinism, 149
Man-nature relationship, 173, 208
Man-nature-society relationship, 23
Mass mobilization, 3
Masters of nature, 112, 121
Material culture, 4, 13
Mineral products, 128–29
Mobilization of resources, 7–8
Mobilizational efforts, 3, 8, 13–14
Modified Area Development Approach (MADA), 20
Money-oriented economy, 61–62
Moral obligation, 157, 214
Morphology, 201
Mother Nature, 147–49, 155, 194–95
Musical instruments, 142, 161–62

Naga societies, 35
Naming ceremonies, 48, 161–62
National Afforestation and Eco-Development Board (NAEDB), 91
National Afforestation Programme (NAP), 91–92, 95
National parks, 51, 74, 106
Nationalization of forests, 75
Natural abode, 49
Natural calamities, 25
Natural cycle, 238
Natural environment, 3, 37, 60–62, 101, 111, 155, 169, 199, 210–11, 215
Natural forces, 146, 157, 194
Natural forest regeneration, 85
Natural gas, 37
Natural hazards, 59
Natural heritage, 62, 71
Natural museums, 189
Natural resource conservation, 44, 104
Natural resource management, 36, 38, 40, 44, 50, 61, 211
Natural resources
community-based, 207
exploitation of, 3, 9
renewable, 10
Nature worship, 27
Nature-culture dynamics, 5
New Social Movements (NSMs), 7–8
Nokrek Biosphere Reserve, 106–7
Non-forest activities, 50
Non-Governmental Organizations (NGOs), 8, 38–39, 68, 71, 107, 242

Non-Timber Forest Products (NTFPs), 36, 43, 47, 49–50, 53–54, 76, 94

Objective analysis, 197
Oral infection, 169
Oral-tradition, 156
Overexploitation, 62, 81, 102, 231

Panicle Initiation (PI), 238
Pielou's evenness index, 78
Pluralism, 18
Political economy, 9
Political life, 17, 60, 62
Polygamy, 138
Protection of Civil Rights Act and Prevention of Atrocities, 21, 24
Protection of nature, 115, 211

Qualitative analysis, 157
Quality of life, 26, 191, 238

Rain-forest, 80
Rainy days, 236–37
Rationalism, 217
Red Panda, 106–7
Religious
 beliefs, 18, 104, 213
 life, 46, 48, 60, 62
 rites, 46, 141, 157–59, 161–62
Religiousness, 118, 122
Reserved Forests, 51
Resource conservation, 43, 209
Resource management, 35, 39, 43
Rice-beer, 140, 142, 149, 160
Rites, socio-religious, 139, 160, 162
Rituals, 28, 48, 137, 147–48, 157, 160, 162–63, 173–74, 180, 183, 209–10
River basins, 234, 236
River systems, 59, 154, 232, 234
River valley projects, 11–12

Sacred forests, 27–29, 76, 105, 194, 213
Sacred groves, 76, 78, 80–81, 104–5, 107, 110, 157, 199–200, 212–13
Sacred mystery, 152
Sacred space, 212–13
Sacredness, 146, 152, 158, 185
Samanvit Gram Vanikaran Samriddhi Yojana (SGVSY), 91
SC/ST Act, 24
Scientific approach, 144–45
Scientific knowledge, 210, 215–18
Seasonal crops, 146–47
Self-sufficient economy, 178–79
Shannon-Wiener index of diversity, 78
Shifting cultivation, 76, 87, 108, 139–40, 171, 237
Siju Bird Sanctuary, 107
Siju forests, 164
Simpson dominance index, 78
Simsang River, 107
Social life, 138, 155, 158, 183
Social movements, 7–8, 43
Social relation, 38, 40
Societies Registration Act, 91
Socio-cultural life, 11, 182
Socio-cultural worldview, 11
Socio-economic life, 48, 127, 132
Socio-religious life, 6
Soil
 conservation, 226
 erosion, 12, 67, 102, 105, 108, 226
 fertile, 67, 154
 red earthen, 160
South Garo Hills District, 107
Spatio-temporal accessibility, 231
Species
 ape, 107
 bamboo, 108, 161, 226
 co-dominant, 80
 dicotyledonous, 164
 diverse, 106
 dominant, 80
 endangered, 11, 64, 110
 endemic, 29, 85

endemic forest, 104
exotic, 28
fauna, 106
fodder, 226
forest, 75
grass, 224
living, 154
plant, 37, 75, 78, 223
richness, 75, 77, 79, 81, 85–86
threatened, 80, 86, 110
tree, 75, 86, 109
woody, 77, 79–80, 220
Spiritual beliefs, 212, 217
Spirituality, 213
Sport, traditional, 64
Standard deviation, 116, 118
Staple food, 130, 140, 142
State Forest Development Agencies (SFDAs), 91
Sub-Tropical Pine Forest, 164
Superstitious beliefs, 176–77
Surface water resources, 234
Sustainability
ecological, 25, 27, 47
environmental, 101, 106, 212, 217
Sustainable development, 38, 101–2, 107, 110
Symbiotic relationship, 6, 27, 194, 199, 213, 217
Symbolic nature, 209

Teacher-trainees, 114–18, 120–22
Timber extraction, 36–37, 78–79, 86
Tobacco, 109, 128, 130, 156
Tooth ache, 167
Tourist guide, 71–72
Traditional
beliefs, 81, 107, 115, 156, 183
communities, 27
craft, 200
cultivations, 108
culture, 170
dresses, 141
Drum, 166
ecological knowledge, 71, 207–13, 215–19
ecological knowledge, 71, 207–9
knowledge, 108, 156, 167–69, 197, 199–200, 202, 209, 215–16
lifestyles, 155, 211–12
practices, 27, 61–62, 70–71, 103, 138, 157, 163
Traditions
abstract, 216
oral, 48, 157, 211
positivist, 216
revive, 40
Transformative changes, 7–8
Tribal
belief, 146
culture, 168–69, 215
development, 19–20, 23–24, 215
population, 18–20, 52
sub-plan, 20
worldview, 215
Tribes Advisory Council, 24
Tropical Forest, 164

UN Water, 231, 239
UNESCO, 106–7, 113–14
United Nations, 16–17, 26, 102, 113
United Nations Agenda, 102
United Nations Millennium Declaration, 101
Unity of nature, 183, 194
Universal Declaration of Human Rights, 16
Universal moral rights, 16
Urbanization, 3, 12–13, 75–76, 178, 180, 231
Utensils, 149, 162
copper, 128

Values
aromatic, 66
articulated, 157
eco-centric, 214
economic, 38

indices, 79
instrumental, 112
personal, 114
spiritual, 47, 208
utilitarian, 158
Veneer factories, 12
Vernal Wood, 102
Village
administration, 36, 52, 54, 224, 226
community, 8, 39, 49, 90, 226
council, 37, 50, 64, 66, 138
development, 9, 97
durbars, 46, 225
environment, 70
forest development committee, 90–98
infrastructure, 71
leaders, 65
streets, 67–68, 71
vicinity, 64

Wage
earner, 96
employment, 53
labour, 97
Waste littering, 68
Water
agricultural, 232, 234
bodies, 11, 232, 234, 238, 240, 242
clean, 67
drinkable, 237
irrigational, 231
quality, 231, 234
resources, 231, 234
security, 231–32, 234, 239, 241–42
surface, 231
trans-boarder, 232
vulnerability index, 232
Welfare, 23, 28
economic, 19
human, 191
tribal, 20
West Garo Hills, 106
Wet places, 199
Wet rice cultivation, 148–49
Widow remarriage, 138
Wild
animals, 36, 60, 65, 140, 184
apples, 66
boar, 37
elephants, 106
games, 63–64
vegetables, 81
Wildlife, 43, 62, 64, 74, 107, 192
Wildlife Protection Act, 81
Winter rainfall, 240
Women activists, 44
World Commission on Environment and Development, 26
World Conservation and Monitoring Centre, 75
World heritage criteria, 107
World's heaviest rainfall region, 232
Worship, 48, 104, 138, 147, 173

Youth Organization, 65

Zabo practice, 30
Zeroline, 131–32